ENCYCLOPAEDIA OF HELMINTHES

ENCYCLOPAEDIA
OF
HELMINTHES

Vol. 2

Aschelminthes

By

Arvind N. Shukla

Rajiv Tyagi

ANMOL PUBLICATIONS PVT. LTD.

NEW DELHI - 110 002 (INDIA)

ANMOL PUBLICATIONS PVT. LTD.
4374/4B, Ansari Road, Daryaganj
New Delhi - 110 002
www.anmolbooks.com

Encyclopaedia of Helminthes

First Edition, 2002

ISBN 81-261-1043-0 (Set)

PRINTED IN INDIA

Published by J.L. Kumar for Anmol Publications Pvt. Ltd., New Delhi - 110 002 and Printed at Mehra Offset Press, Delhi.

Contents

Preface

Helminthes constitute a large assemblage of *worms* of comparatively simple and varied organization. Helminthes is derived from the Greek word *helmins* means *worms*. The term worms is not correct because it is applied to elongated invertebrates without appendages and with bilateral symmetary. In 18th Century all the worm-like animals were included in an old phylum- *Vermes* (*Linnaneus*, 1735). But this was a very heterogenous group which included a highly diverse assemblage of forms. In the modern classification this group is divided into several independent phyla such as Platyhelminthes, Aschelminthes, Annelida etc.

The term helminthes is restricted to a few phyla of invertebrate animals, all of which are superficially worm like but they differ markedly in their morphology, life-history and bionomics. The helminthes are restricted to two phyla of Animal Kingdom. These are:

1. Platyhelminthes or flat-worms.
2. Aschelminthes or round-worms.

The present work is designed to present all vital information on Helminthes. The entire information is gathered in three volumes namely:

- Platy Helminthes
- Aschelminthes
- Physiology of Helminthes

Profusely illustrated in the light of latest researches, this work will prove a dependable reference work for students, teachers and researchers in the discipline.

—**Authors**

ASCHELMINTHES

Chapter—1

General Character & Classification of Aschelminthes

The nematodes are vermiform Aschelminthes of cylindrical shape, without cilia or flame bulbs, with four main longitudinal epidermal chords, a three-angled pharynx, circumenteric nerve ring, copulatory spicules, and one or two tubular gonads opening separately in the female, into the rectum in the male.

External Characters.—The nematodes, or roundworms, are vermiform animals of long cylindrical shape, circular in cross section. There are two general types of body form, the *fusiform* and the *filiform*. The fusiform shape is that of an elongated spindle, widest through the middle and tapering toward the blunt or pointed ends; the posterior end is generally more tapering and pointed than the anterior end and in some species as *Rhabditis filiformis* is very slender. In the filiform type, less common than the fusiform but illustrated in the Mermithidae, the Filarioidea, and the trichuroid genus *Capillaria*, the body is thread-like and of uniform diameter throughout, not diminishing toward the ends. Other variations are the short, plump, pyriform or oval shape assumed by certain ripe females as the phytoparasitic *Heterodera*, and the trichurine type in which the body is filiform anteriorly, fusiform posteriorly. The posterior end of the two sexes frequently differs in shape

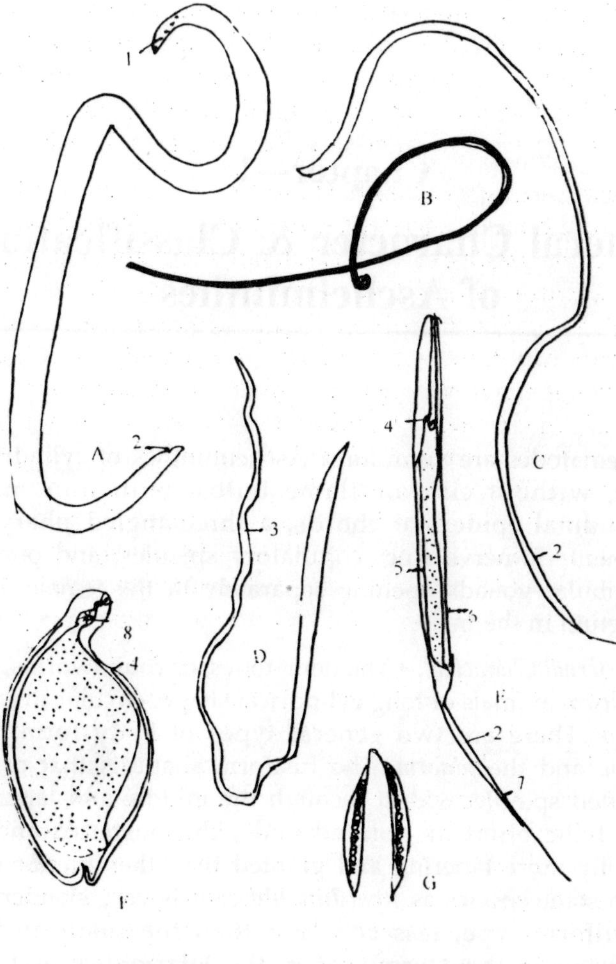

Fig. 1.1 : Types of nematodes. *A*. Fusiform type, dog ascarid, *Toxascaris leonina*, original. *B, C*. Filiform type, *Capillaria brevis B, male; C, female D. Trichuris trichiura* partly fusiform, original. *E. Rhabditis filiformis* with slender tail. *F. Heterodera schachtii*, mature female. *G*. Anterior end of *A*, showing cervical alae. 1, cervical alae; 2, anus; 3, anterior slender part; 4 pharynx; 5, intestine; 6, vulva; 7, tail; 8, spear.

and external features, being generally curved in the male, and provided with papillae, alae, etc., lacking in the female.

The free-living nematodes are generally microscopic or very small, below 1 mm. in length; among fresh-water and terrestrial forms the maximum length is several millimetres, but the marine nematodes, which are in general the largest of the free-living members, may attain a length up to 5 mm. While some parasitic nematodes are also of small size, many reach considerable length and some are excessively long. Thus the females of the kidney worm, *Dioctophyme renale,* and the guinea worm, *Dracunculus medinensis,* may attain a length of a metre or more. The males are nearly always smaller than the females and this size difference between the sexes is quite marked in some species.

The nematodes in general lack coloration, being transparent or of a whitish or yellowish tint conferred on them by the cuticle.

The body is not divisible into definite regions and lacks a distinct head although this term is sometimes applied to the anterior end to a level just behind the nerve ring. In the marine family Draconematidae the swollen anterior end simulates a head but represents the entire pharyngeal region. The ventral surface of nematodes is identifiable by the presence in the mid-ventral line of the excretory pore, the gonopore in the female, and the anus. The excretory pore has an anterior position. The female gonopore, also called *vulva,* is usually situated in the posterior body half but may occur anywhere in the mid-ventral line. The anus, which also serves as male gonopore, lies near the posterior end, and the body region behind it or postanal region, is commonly called *tail.*

The mouth occupies the center of the anterior tip and is encircled by structures showing a pronounced hexamerous or biradial arrangement. Only in recent years through the study of marine nematodes by Filipjev, de Coninck, Stekhoven, and others, has an understanding of the generalized morphology of the nematode head been attained. The views especially of de Coninck (1942) are here adopted. The work of these men has shown that the mouth is primitively surrounded by six lobes, the *lips* or *labia*, each of which bears a sensory papilla, the *inner labial papilla*, sometimes altered to a bristle; the lips are laterally located, three on each side. The area bearing the lips may be separated from the remainder of the anterior tip by a circular groove. Outside the inner labial circle of six sense organs is a second or *outer labial* circle of six papillae or bristles, which, according to de Coninck, also belong to the lips but are regarded as cephalic by others. This is followed by an outer

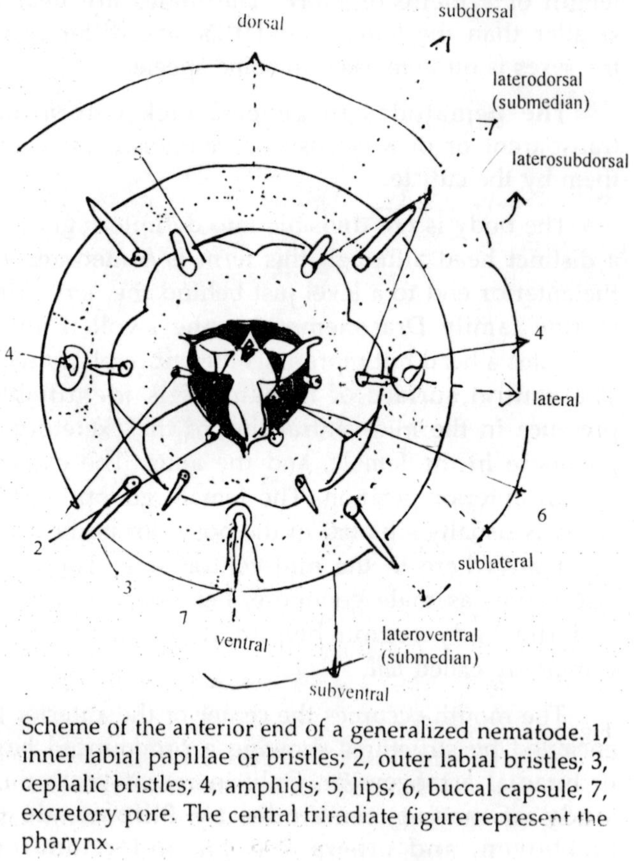

Fig. 1.2 : Scheme of the anterior end of a generalized nematode. 1, inner labial papillae or bristles; 2, outer labial bristles; 3, cephalic bristles; 4, amphids; 5, lips; 6, buccal capsule; 7, excretory pore. The central triradiate figure represent the pharynx.

circle of four bristles or papillae which lie outside the labial region, and hence are termed *cephalic*. There are thus recognized as typical and primitive parts of the nematode head three circlets or crowns of sense organs of the nature of bristles or papillae, to a total number of 16. This condition obtains in many marine nematodes but several variants occur. The outer or cephalic crown of four bristles may move forward and join the outer labial crown of six bristles, making one circlet of ten bristles, the compound nature of which, however, is usually indicated by the differing size of the two components, as the cephalic bristles are usually shorter than the outer labial bristles. The number of bristles

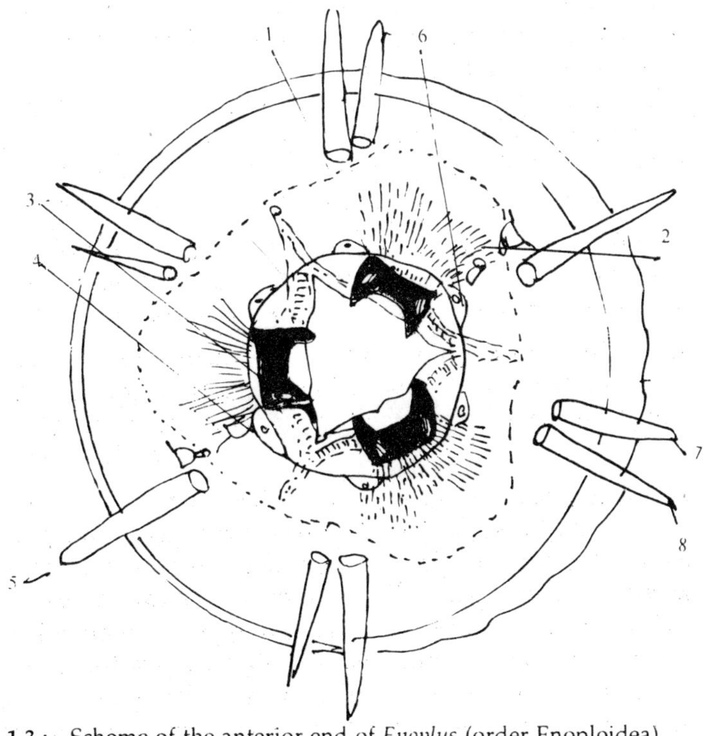

Fig. 1.3 : Scheme of the anterior end of *Enoplus* (order Enoploidea), showing two outer circlets of bristles combined to one of ten bristles. 1, jaw; 2, jaw, muscles; 3, lips; 4, labial papillae; 5, amphid; 6, cephalic slit; 7, cephalic bristles; 8, outer labial bristles.

is sometime augmented by doubling or by the forward migration of body sensory bristles. On the other hand, the number of head sense organs is often reduced in parasitic and terrestrial nematodes. The lips in these forms may be fused in pairs to form three lips, one dorsai and two ventrolateral, or be reduced to two lateral lips or be altogether wanting. The six inner labial papillae are usually retained, however, so that in such cases there are two or three per lip or else they simply encircle the mouth opening. Small lobes termed *interlabia* are sometimes present between the lips. In parasitic and terrestrial nematodes the sensory bristles of the two outer circlets are usually altered to papillae, the outer labial and the cephalic papillae, and in fact bristles and papillae are to be regarded as interchangeable structures. In such nematodes the two outer circlets are usually concentrated into one circlet and are frequently reduced in number by loss or fusion. There are thus in terrestrial and parasitic nematodes usually only two circlets of sense organs present, which may conveniently be termed the inner or internal and the outer or external circlets. In parasitic nematodes the inner circlet is also often wanting or vestigial (*Ascaris*) although its nerve supply can usually be traced histologically.

From the foregoing it is evident that a marked hexamerous or biradial symmetry pervades the anterior end of primitive nematodes; this symmetry is exhibited by the lips, the two labial circlets of sense organs, and the anterior part of the digestive tract. It is indicative of the superposition of a radial symmetry on an original bilateral symmetry and supports the conclusion of Steiner (1921) that nematodes originally led a semisessile life, fastened to objects by their posterior tip by means of the adhesive secretion of the caudal glands and waving about in the water. The four cephalic sense organs of the outer circlet retain their bilateral symmetry and are the anteriormost members of body sense organs or cuticular structures, which tend to be arranged in four or eight longitudinal rows.

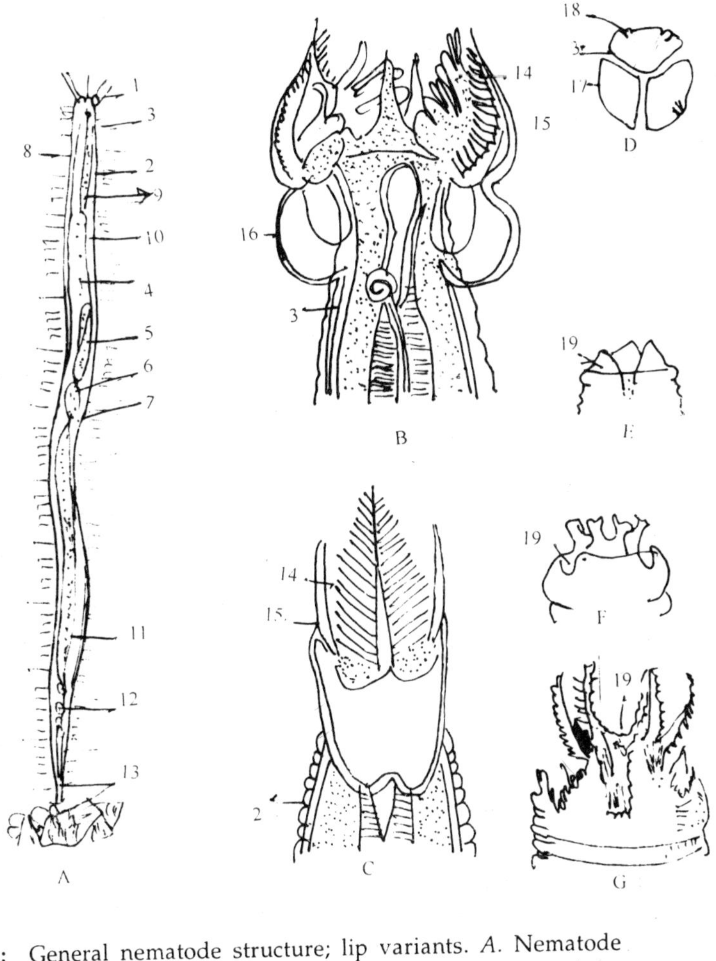

Fig. 1.4 : General nematode structure; lip variants. *A*. Nematode undulating in attached position. *B, C*. Two views of the anterior end of *Wilsonema* (order Araeolamoidea), showing lip excrescences. *D*. Scheme of the lips of *Ascaris*. *E-G*. Types of probolae of *Acrobeles*. 1, head bristles; 2, pharynx; 3, amphid; 4, intestine; 5, ovary; 6, uterus; 7, vulva; 8, nerve ring; 9, excretory pore; 10, renette; 11, anus; 12, caudal adhesive glands; 13, adhesive tube; 14, dorsal and ventral excrescences; 15, lateral excrescences; 16, buccal capsule; 17, outer labial papillae; 18, cephalic papillae; 19, probolae.

It is notable that none of the lips or head sense organs occupy medial positions. Their positions may be designated in a general way as dorsolateral, lateral, and ventrolateral. In the cephalic circlet it is the lateral sense organs that are absent and it seems probable that these were primitively wanting although some nematodes exist in which there are six sense organs in the cephalic circlet. To indicate the exact position of parts in the head circlets an elaborate terminology has been devised by some nematodologists as the Chitwoods.

The lips not infrequently bear or are surrounded by various cuticular protuberances and excrescences. In certain terrestrial nematodes as *Acrobeles* and *Wilsonema* peculiar rigid lip excrescences termed *probolae,* three or six in number, project forward; they vary from simple rounded, conical, or forked eminences to branched projections resembling antlers. When six are present they are arranged in two circlets of three each. Their function is uncertain; possibly they serve to shove the earth aside or in feeding but can be moved only by lip motions. In many Strongylidae, the lips are altered into an upstanding collar, the inner surface of which is subdivided into a few to many (40 or more) lobes or teeth termed the *leaf crown* (or *corona radiata*), accompanied in a number of genera by a similar *inner leaf crown*.

Various other cuticular specializations occur on the anterior end outside the lips, especially in the order Spiruroidea. Thus in the spirurine nematodes, the lips are encircled by a cuticular collar that is most prominent dorsally and ventrally where it may be extended into conspicuous *head shields* that arch over the lips. Several genera of the spiruroid family Physalopteridae also possess an upstanding cuticular collar or collaret around the anterior extremity that in some genera can withdraw into it. In still another spiruroid family, the Acuariidae, there are present anteriorly two, four, or six longitudinal cuticular cord-like thickenings or grooves termed *cordons* or *epaulets*. These may be straight, recurved, or form loops or horseshoes; in *Seuratia,* the free posterior edge of the four cordon scallops is formed into

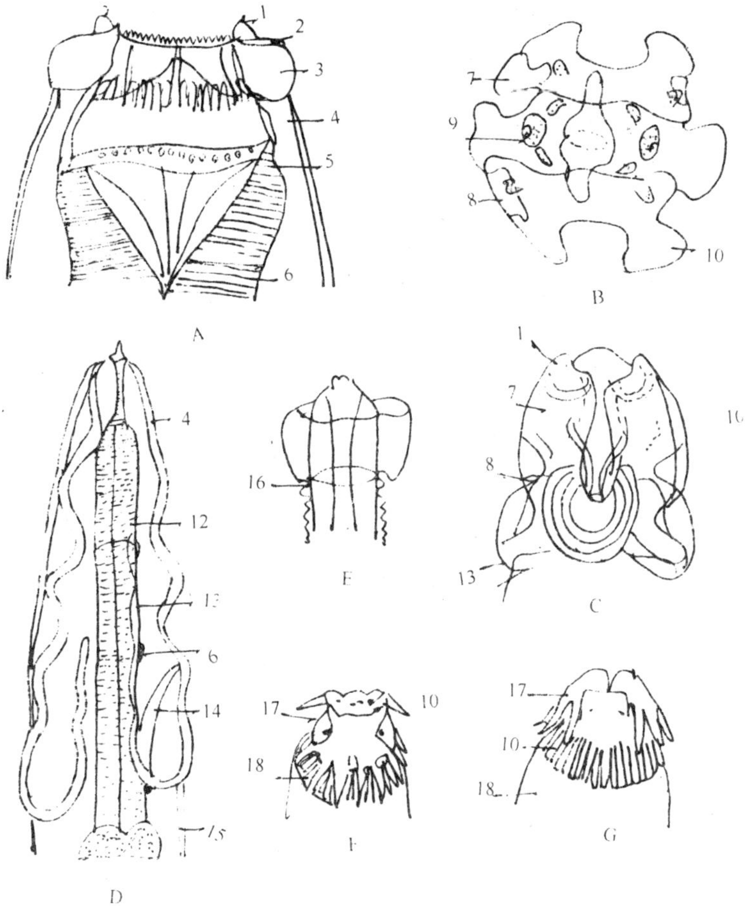

Fig. 1.5 : Head structures. *A*. Horse strongyle, *Poteriostomum,* with leaf crowns, lateral view. *B*, C. Spiruroid *Parabronema,* from elephants, with head shields and cordons. *B. End face* view. C. Lateral view. *D. Dispharynx,* bird acuariid, with recurved cordons. *E. Proleptus,* shark spiruroid, with cuticular collar. *F*, *G*. Two views of the head of *Histiocephalus*. 1, lip papilla; 2, outer leaf crown; 3, inner leaf crown; 4, buccal capsule; 5, dorsal gutter (carries gland duct); 6, pharynx; 7, head shield; 8, lateral lips; 9, amphid; 10, external circlet of papillae; 11, mouth; 12, nerve ring; 13, cordon; 14, renette; 15, intestine; 16, collar; 17, head appendages; 18, cervical collar.

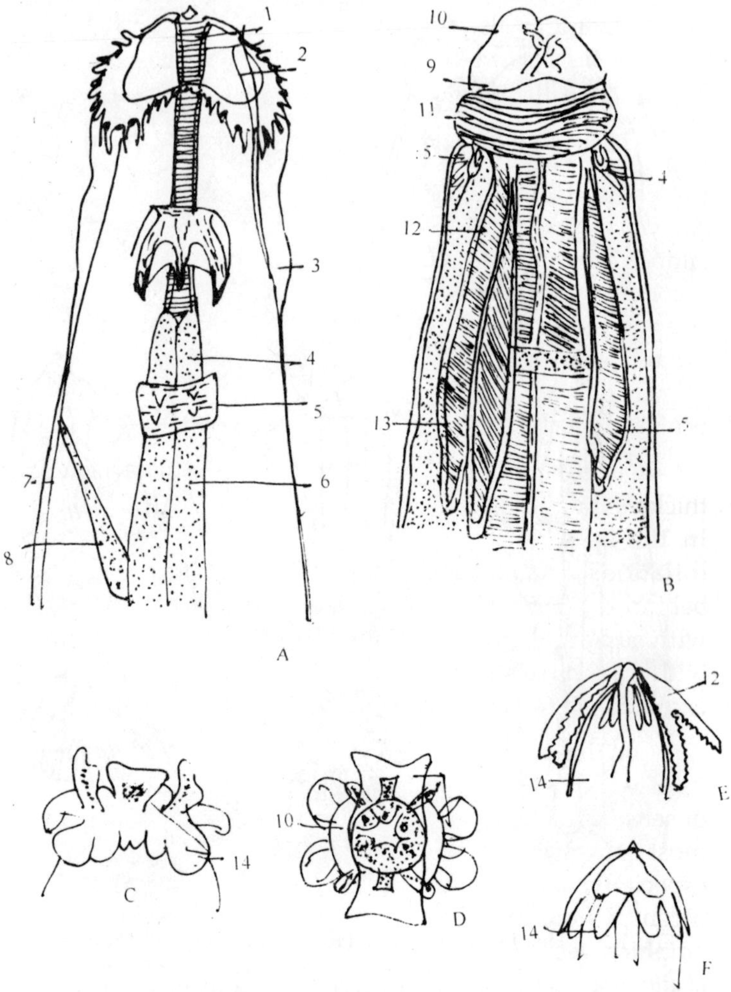

Fig. 1.6 : Head structures (continued). *A. Seuratia* with spiny cordons and body spines. *B. Tanqua* with ballonets and cervical sacs. *C, D.* Two views of the head end of *Serticeps*. *E.* Anterior end of *Ancyracanthus*. *F.* Anterior end of *Ancyracanthopsis*. 1, buccal capsule; 2, cordon; 3, trident spine; 4, pharynx; 5, nerve ring; 6, rows of body spines; 7, excretory pore; 8, renette; 9, lip; 10, labial papilla; 11, head bulb.; 12, cervical sacs; 13, cervical papillae; 14, head appendages; 15, collar.

spines. Cordons also occur in the oxyuroid genera *Aspidodera* and *Pseudaspidodera*. In the ascaroid genera *Heterocheilus* and *Typhlophorus* there is a broad cuticular thickening behind the lips that is longitudinally ribbed in the latter. In several spiruroid genera there are fantastic cuticular ornamentations projecting from the head as four feathery appendages in *Ancyracanthus,* four pointed wings in *Schistorophus,* eight simple lobes in *Ancyracanthopsis,* two split into a number of secondary lobes in *Histiocephalus,* and a circle of variously shaped processes in *Serticeps*. Mention of other head ornamentations in spiruroids will be found under the order Spiruroidea in the discussions of the families Acuariidae, Schistorophidae, and Histiocephalidae.

It is common for the cuticle of the anterior end to be thickened or inflated through the presence of vesicles therein. In the spiruroid family Gnathostomidae, four cuticular inflations termed *ballonets* result in a swollen band just behind the lips called the *head bulb*. The head bulb is armed with circlets of spines in *Gnathostoma* and *Echinocephalus* and with transverse striations in *Tanqua* in which it is subdivided externally into two or four swellings.

On the sides of the anterior end, just external to or level with the circlet of cephalic sense organs, occurs a pair of sense organs very characteristic of nematodes, the *amphids,* most conspicuous and best developed in aquatic nematodes, especially marine ones, but also present in a reduced and inconspicuous state in terrestrial and parasitic members. The amphids are cuticular excavations of three general shapes: *cyathiform, spiral,* and *circular*. The cyathiform type, found in the Enoploidea, has the shape of a pocket and is provided with a slit aperture. The spiral type, characteristic of the Chromodoroidea and Araeolaimodea, occurs in many variants, ranging from a spiral of several turns to crescentic and loop shapes. The circular amphid has the form of a disk and is limited to the order Monhysteroidea. Students of free-living nematodes (Filipjev, de Coninck, and Stekhoven) have utilized the shape of the amphids as a basis for classifying nematodes and as their system appears to be the most satisfactory, it has been adopted here. The amphids are presumably chemoreceptors and are provided with a gland and nerve endings. In addition to and adjacent to the amphids, the Enoploidea have another pair of pouch-like cephalic sense organs, the *cephalic slits*.

The general body surface may be smooth, especially in parasitic members, but often presents a variety of cuticular specializations such as bristles, spines, warts, papillae, punctations, striations, and ridges. Cuticular sensory bristles are often scattered irregularly and sparsely along the body length in marine nematodes. In *Greeffiella* closely placed circles of bristles cover the entire body, giving the animal a very bristly appearance. In the marine families Epsilonematidae and Draconematidae the posterior part of the body bears a double row of large stiff hollow bristles, known as *stilt bristles* or *ambulatory setae*, employed in locomotion.

Transverse or longitudinal striations, thickenings, or ridges are of common occurrence. In many nematodes fine transverse grooves, termed *striations*, occur at regular intervals; when these are deeper and more pronounced they are called *annulations*, and give the body a segmented appearance, dividing the cuticle into rings or *annules*. Such annulation is very conspicuous in the marine families Epsilonematidae, Draconematidae, and Desmoscolecidae. In the last family thick cuticular rings alternate with depressed rings so that the worm greatly resembles an insect larva. Strongly annulated bodies also occur in the phytoparasitic criconematines. Annules may be stiffened in whole or in part by longitudinal flutings or rod-shaped or V-shaped thickenings. Incomplete or partial annules sometimes occur. Annules may be of trapezoidal form, with posterior edges that project over succeeding annules (*Thelazia*,) and these edges may be spiny as in the parasitic genera *Goezia* and *Spinitectus*. In *Criconema* (= *Iota, Ogma*), the rear edge of each annule bears eight scales or in one species transverse rods that project over the succeeding annule and render the body very rigid, capable of motion only in the forward direction. In the curious marine family. Epsilonematidae, ventral combs

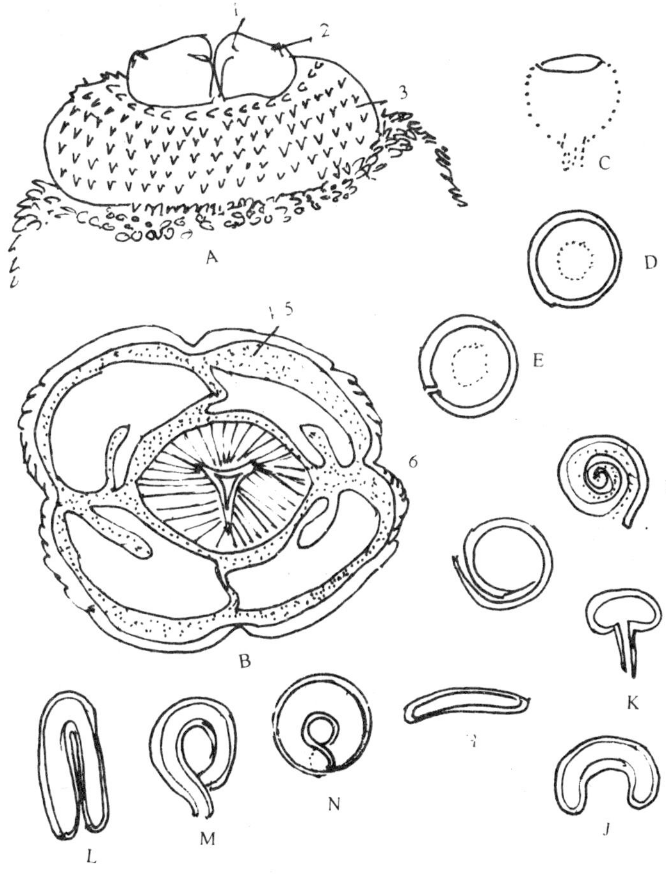

Fig. 1.7 : Head structures (continued). *A. Gnathostoma. B.* Section through the head bulb of *Tanqua,* showing the ballonets. *C–N.* Types of amphids. *C.* Cyathiform type, Enoploidea. *D.* Circular type, Monhysteroidea. *E.* Variant of the circular type, transitional from spiral type (*Linhomeus*). *F.* Spiral type, Chromadoroidea. *G–N.* Variants of the spiral type. *G–K.* Chromadoroidea. *K. Plectus. L–N.* Araeolamoidea. 1, lip; 2, labial papilla; 3, head bulb; 4, ballonet cavity; 5, pharynx; 6, muscles.

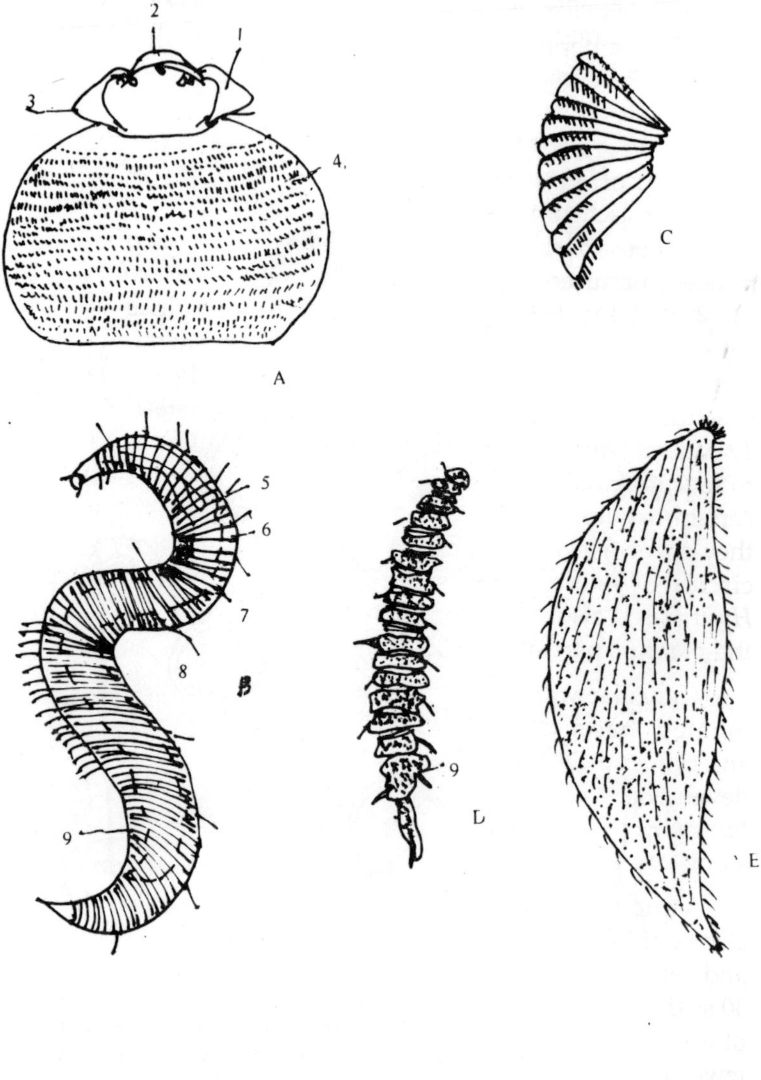

Fig. 1.8 : Cuticular structures (continued). *A. Echinocephalus. B.* An epsilonematid, from life, Puget Sound. *C.* Ventral bristle combs of an epsilonematid. *D. Desmoscolex. E. Greeffiella.* 1, lip; 2, labial papillae; 3, papillae of outer circlet; 4, head bulb; 5, pharynx; 6, end bulb of pharynx; 7, intestine; 8, stilt bristles; 9, anus.

of spines are borne on the rear edges of the annules in the region just anterior to that provided with stilt bristles. The annules are cuticular thickenings that do not extend to interior structures; they may be solid or hollow or filled with a spongy or vacuolated material. They represent the same kind of superficial segmentation that occurs in rotifers and other aschelminths.

Spines apart from those already mentioned are of frequent occurrence, especially in the order Spiruroidea, as illustrated by: *Echinonema* with circlets of spines on the anterior part of the body, *Pneumonema* with lateral rows of large hooks anteriorly; *Gnathostoma* with spines covering practically the entire body but diminishing posteriorly; and *Tetrameres, Rictularia, Rictularioides,* and *Echinuria* with two to four longitudinal rows of spines. Spines are also found in certain oxyuroids as in *Hystrignathus* and *Lepidonema* where they cover the anterior end of females and in *Carnoya,* with circles of spines anteriorly. The dioctophymoid genus *Hystrichis* is also armed with small spines over the whole or the anterior part of the body.

Cuticular warts are sometimes present. In *Gongylonema,* inflated warts, also called plaques, are strewn over the anterior end, in some species on one side only. The best development of warts is seen in the curious terrestrial genus *Bunonema* which has a single or double row of cuticular warts along one side of the body.

Various types of longitudinal cuticular specializations also occur. In many trichostrongyles, as *Nematodirus, Cooperia,* and related genera the body is longitudinally ribbed; about 40 such ridges occur in *Oswaldocruzia leidyi* from the intestine of a tree frog. Usually, however, the longitudinal ridges are fewer in number and higher and more pronounced, and are then termed *wings* or *alae.* Alae occur as *lateral* or *longitudinal* alae, one, two, or four in number usually extending the body length in lateral positions; as *cervical* alae, single or paired, on the anterior part of the body, as in the ascarid genera *Toxocara* and *Toxascaris* and as caudal alae on the rear end of

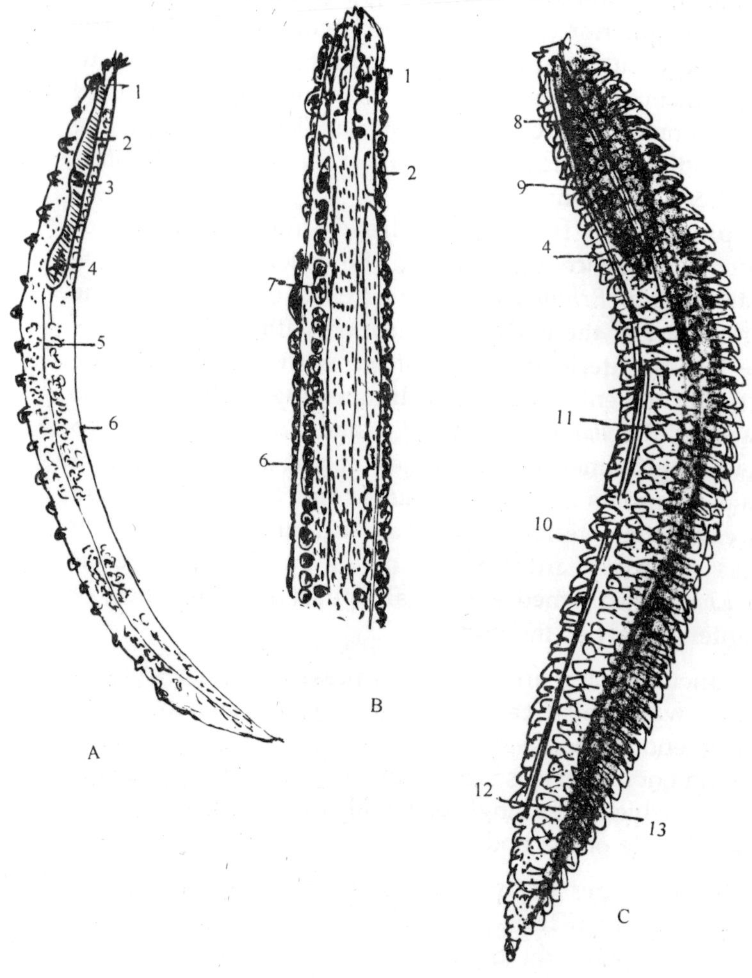

Fig. 1.9 : Cuticular structures (continued). *A. Bunonema* with lateral warts. *B*. Anterior end of *Gongylonema,* with placques. *C. Criconema*. 1, buccal capsule; 2, pharynx; 3, nerve ring; 4, end bulb; 5, lateral chord; 6, warts; 7, excretory pore; 8, buccal stylet; 9, muscles of stylet; 10, ovary; 11, intestine; 12, valve 13, anus.

males. Caudal alae, although laterally situated, are not part of the lateral alae. They are employed in copulation and hence generally bear genital papillae. They are usually absent from marine nematodes but are of common occurrence among saprophagous and zooparasitic nematodes where they vary from narrow ridges to wide expansions, sometimes confluent ventrally in front of the anus. They are usually purely cuticular but in the Strongyloidea the wide caudal alae, meeting posterior to the tail to form a copulatory bursa, are provided with characteristic muscle rays.

Finally mention may be made of a type of cuticular marking termed *punctation*. Punctations are minute depressed dots or ovals; they may occur in transverse or longitudinal rows and often are arranged in patterns. Sometimes transverse striations are found on higher magnification really to consist of rows of punctations; punctations may also occur between striations or in a row along ridges.

Papillae, presumably always of sensory nature, are among the common surface structures of nematodes. In addition to the labial and cephalic papillae already discussed the anterior region frequently bears in the vicinity of the nerve ring a pair of *cervical* papillae. These have been named *deirids* by Cobb but no reason is apparent for distinguishing them from other papillae by a special name. In *Chevreuxia*, the cervical papillae are covered by a backwardly directed cuticular fold. Various types of *genital* papillae, associated with copulation, occur on the ventral surface of the rear end of males. The ordinary genital papillae are a few to many small rounded cuticular elevations, arranged irregularly or in longitudinal rows, in both preanal and postanal locations, as well as on the caudal alae when present. They may be elevated on little stalks and are then termed *pedunculated* papillae. They are sometimes encircled by punctations forming a *rosette* or supported fore and aft by little cross-striated cuticular plates called *plectanes*. Both rosettes and plectanes occur in the oxyuroid family

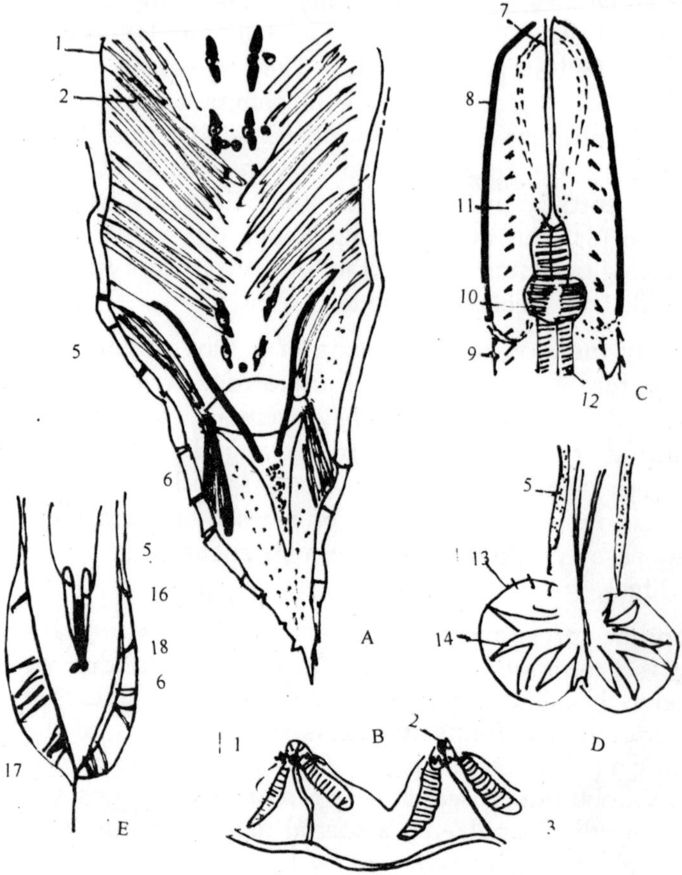

Fig. 1.10 : Cuticular structures, copulatory aids. *A*. Rear end of *Cosmocerca,* with plectanes. *B*. Two plectanes, enlarged, side view. C. Anterior end of *Echinuria* with cordons and four rows of spines. *D*. Strongyloid bursa (*Longistriata*). *E*. Rear end of *Rhabditis maupasi,* with caudal alae and pedunculated papillae. 1, papilla; 2, frill around papilla; 3, plectane; 4, muscles; 5, spicules; 6, ordinary genital papillae; 7, lips; 8, buccal capsule; 9, pharynx; 10, nerve ring; 11, cordon; 12, rows of spines; 13, bursa; 14, muscle rays of bursa; 15, nerve to papilla; 16, caudal alae; 17, pedunculated papillae; 18, gubernaculum.

Cosmocercidae. Various other papilliform copulatory aids are found near the anus in male nematodes. There may be a preanal mid-ventral row of *copulatory warts,* larger elevations

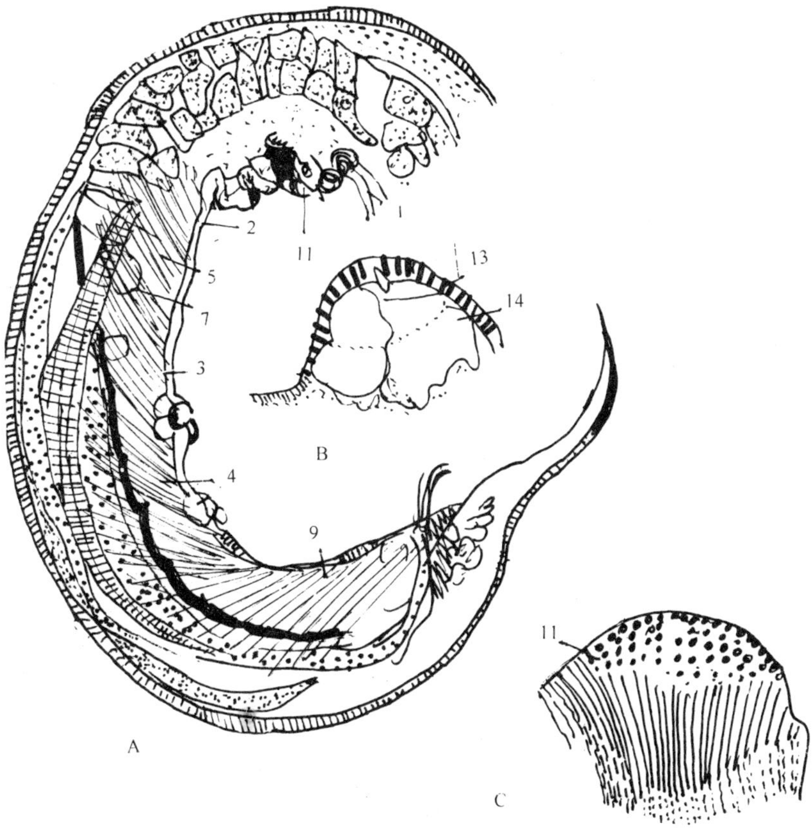

Fig. 1.11 : Copulatory aids (continued). *A.* Rear end of *Cosmocercella haberi* with rosetted copulatory warts. *B.* Section through a wart. *C.* Enlarged view of a wart. 1-4, four pairs of copulatory warts; 5, sperm duct; 6, intestine; 7, spicule retractor; 8, spicules; 9, caudal alae; 10, ordinary genital papillae; 11, punctations of rosette; 12, depressor ani muscle; 13, sensory organ; 14, nerve.

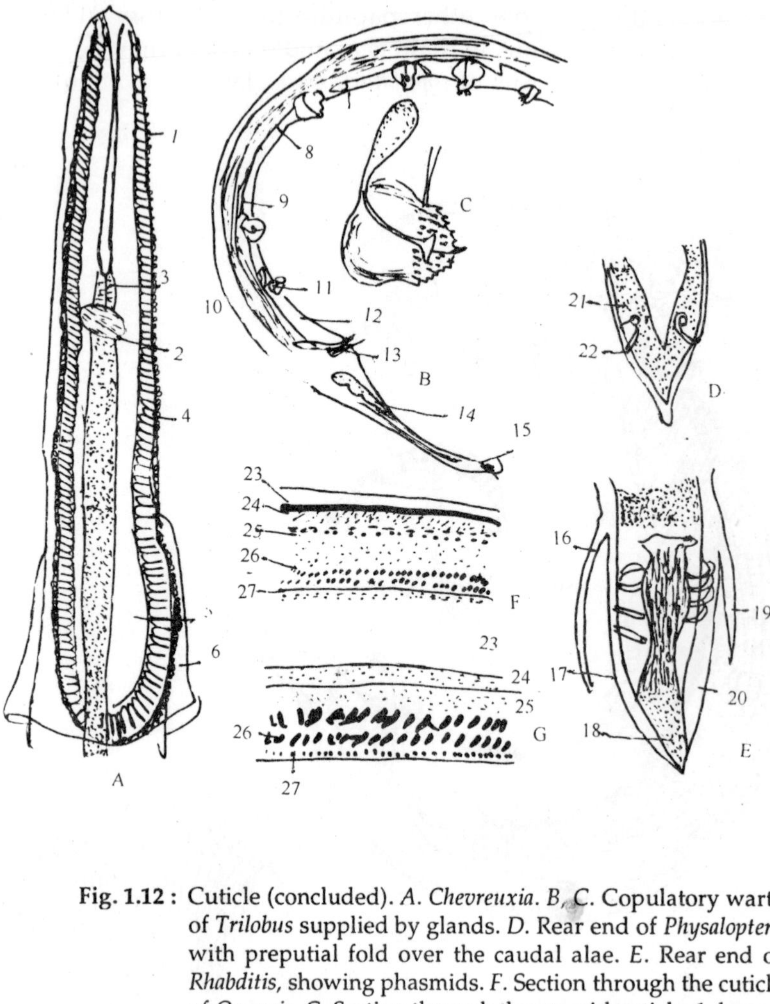

Fig. 1.12 : Cuticle (concluded). *A. Chevreuxia*. *B, C.* Copulatory warts of *Trilobus* supplied by glands. *D.* Rear end of *Physaloptera* with preputial fold over the caudal alae. *E.* Rear end of *Rhabditis,* showing phasmids. *F.* Section through the cuticle of *Oxyuris*. *G.* Section through the ascarid cuticle. 1, buccal capsule; 2, pharynx; 3, nerve ring; 4, cordon; 5, cervical papilla; 6, cuticular fold; 7, copulatory wart; 8, gland of same; 9, seminal vesicle; 10, intestine; 11, spicule; 12, gubernaculum; 13, caudal glands; 14, their duct; 15, adhesive tube; 16, anus; 17, pedunculated papillae; 18, ordinary papillae; 19, preputial fold; 20, caudal alae; 21, gland of phasmid; 22, duct of phasmid; 23, cortical layer; 24, fibril layer; 25, matrix; 26, fiber layers; 27, basement membrane.

with a central concavity. A single large sucker-like papilla, the *preanal sucker,* occurs in the mid-ventral line shortly anterior to the anus in several families of zooparasitic nematodes. In the oxyuroid family Heterakidae, this sucker has a definite cuticularized rim, lacking in the preanal suckers of the oxyuroid families Subuluridae and Kathlaniidae and the spiruroid family Cucullanidae. The sucker may be circular or oval and is supplied by muscle fibres.

There are still other caudal cuticular specializations in males that presumably act as copulatory aids such as scales and areas provided with transverse or longitudinal ridges. Flask-shaped excavations or cuticular tubes occurring singly or in a preanal row and serving as gland outlets also appear to function in copulation.

Near the posterior end of many nematodes there occurs a pair of cuticular pouches resembling the amphids. These are called *phasmids* by Cobb. They appear normally to be the outlets of the precaudal glands, and a surface papilla is often associated with them. Glands and phasmids may degenerate, leaving only the papilla as evidence of their former existence. Phasmids occur chiefly in the parasitic nematodes. So much importance has been placed on the phasmids by the Chitwoods (1937) that they have divided the Nematoda into two subclasses, Phasmidia with and Aphasmidia without phasmids. As this scheme has not found favor with other nematodologists, it is not adopted in the present book.

At the posterior tip of the majority of free-living nematodes there occurs a pore which is the outlet of the caudal glands (see below). This pore may be mounted on a tube, entirely comparable to the adhesive tubes of gastrotrichs and other aschelminths. This whole adhesive apparatus is identical with that of rotifers and gastrotrichs and its presence constitutes valuable evidence for the relationship of nematodes to these two classes. The apparatus is absent in parasitic nematodes.

In addition to the caudal alae and strongyloid bursa already noted, other caudal terminal enlargements occur. Males of the order Dioctophymoidea also have a terminal bursa but this is formed of an expansion of the entire body wall, not just of the cuticle. In females of the spiruroid genus *Simondsia* the caudal end is greatly expanded into a rounded excrescence containing he distended uterus. In some species of the spiruroid genus *Physaloptera*, the tail is covered by a backwardly directed bell-like fold of cuticle.

The foregoing account by no means exhausts the variations of external forms and cuticular structure but should give some idea of their range.

Body Wall.—The body wall of nematodes consists of cuticle, epidermis (often called subcuticle), and muscle layer. The cuticle, of which the external specializations were recounted above, is of complicated histological and chemical construction. Histologically it consists of several layers the number of which is greater in the larger parasitic members than in the smaller forms but which are reducible to three kinds of material: the cortex, the matrix or homogeneous stratum, and the fiber layers. The cortical layer consists of a dense material of the nature of a keratin and is resistant to solvents and to digestion. The matrix layer commonly has a spongy or finely alveolar appearance and according to Chitwood (1936) consists of or contains a fibroid named matricin, rich in sulphur. The matrix layer is often subdivided by one or more narrow strata of felt-like fibrils and just beneath the cortical layer may be more or less altered to form what is termed the internal cortical layer. The innermost part of the cuticle consists of two or three fiber layers of very dense connective tissue running in different directions in adjacent layers. Like vertebrate connective tissue, these fiber layers consist chiefly of collagen. Through the subdivision of the matrix layer by fibrillar membranes and the presence of the maximum number of fiber layers, as many as eight or nine layers may be counted in the cuticle of ascarids. The cuticle is bounded internally by a basement membrane. According to the chemical investigations of Chitwood (1936), the cuticle of *Ascaris* consists of 35 per cent matricin, 29 per cent collagen, and 2.2 per cent keratin, all reckoned as dry weight. Other substances found that could not be correlated with the histological layers were albumins and a glucoprotein.

The layers of the cuticle participate to varying degrees in the formation of external cuticular structures. Transverse striations and annulations may or may not involve regularly repeated thinnings of the cortical layer. Cuticular outgrowths generally have a core of matrix. The alae consist of matrix covered by the cortical layer and cervical alae contain hard skeletal supports embedded in the matrix.

The epidermis (hypodermis, subcuticle) is a cellular or syneytial layer that bulges into the pseudocoel at four places to form four longitudinal ridges termed the *longitudinal chords,* middorsal, mid-ventral, and lateral in position. Of these the lateral chords are the most conspicuous and usually show on the surface as pale lines. In some nematodes, four subsidiary chords occur between the four main chords, in dorsolateral (subdorsal) and ventrolateral (subventral) positions. The chords are better developed anteriorly and tend to disappear posteriorly, except the lateral chords which commonly run the body length. The epidermal nuclei are situated in the chords and the epidermis between the chords (subcuticle of older authors) is devoid of nuclei. Typically there are three longitudinal rows of nuclei in the lateral chords and zero to three rows in the median chords. The subsidiary chords and often also the dorsal chord lack nuclei but the ventral chord is generally provided with them. Cell walls are usually present in the epidermis. The cells of the central row of the lateral chords are cuboidal but the other epidermal cells are of peculiar shape, consisting of a nucleated bulging part situated in the chord and a flattened nonnucleated extension on one or both sides that forms the epidermis between the chords. The number of cells or nuclei in each longitudinal row in limited in juvenile nematodes

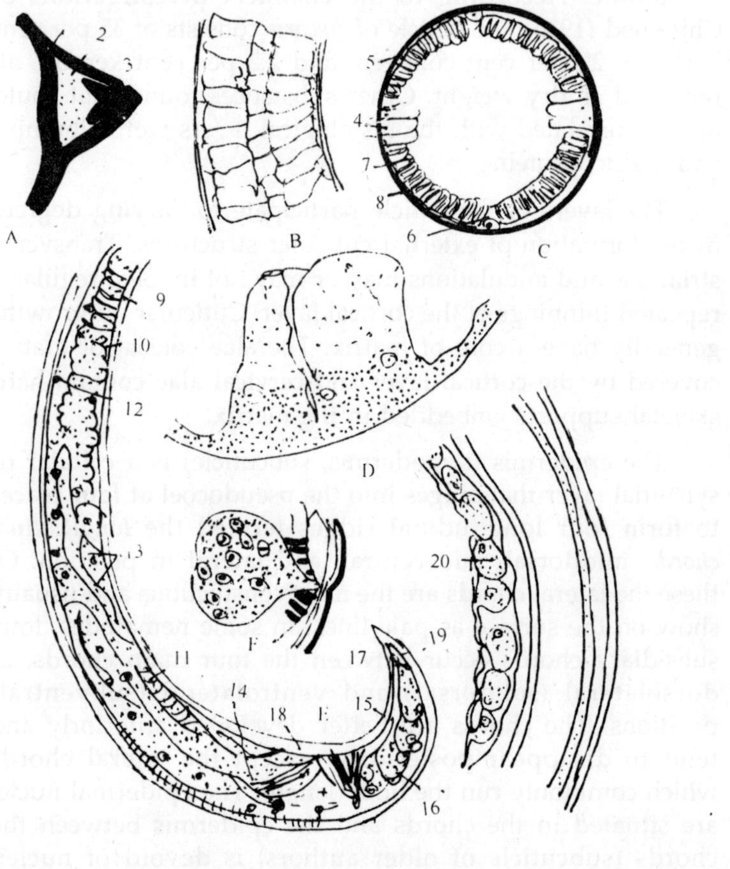

Fig. 1.13 : Body wall, epidermal glands. *A.* Section through the cervical alae of the cat ascarid. *B.* Surface view of the lateral chord, showing outlines of the epidermal cells. *C.* Schematic section through a marine polymyarian nematode. *D.* Section through the lateral chord (*Cuculianus*) *E.* Side view of rear end of male araeolamoid, showing lateral chord glands. *F.* Lateral chord gland enlarged. *G. Axonolaimus,* with mid-ventral row of gland cells 1, cortex; 2, matrix; 3, stiffening; 4, lateral chord; 5, dorsal chord; 6, ventral chord; 7, muscles; 8, cuticle; 9, intestine; 10, testis; 11, seminal vesicle; 12, lateral chord gland cells; 13, copulatory glands; 14, spicule; 15, gubernaculum; 16, muscles of male apparatus; 17, caudal glands; 18, anus; 19, ordinary genital papillae; 20, mid-ventral glands.

and adults of smaller species, ranging from about 12 to 25. In the larger forms the epidermis tends to become syncytial in later life and the nuclei may divide to form large numbers of small nuclei.

In the free-living nematodes, unicellular epidermal glands are of frequent occurrence but are generally lacking in parasitic members. Gland cells usually accompany the amphids and phasmids and other sensory organs (see further below). A single or double row of gland cells situated in the lateral chords and opening through the cuticle has been observed in many marine nematodes of the orders Enoploidea, Chromadoroidea, and Araeolamoidea. Although the ordinary genital papillae lack glandular components, being apparently tangoreceptors, gland cells usually accompany the copulatory warts and other accessory copulatory structures of males. A mid-ventral preanal row of gland cells occurs in males of some species of the araeolamoid genus *Axonolaimus* and a similar row provided with tubular exits is found in several chromadoroid genera. Caudal glands are generally present in the tail of free-living marine nematodes, usually lacking in others. They are long-stalked, pyriform cells, nearly always three in number, that open at the tip of the tail by a single pore, sometimes mounted on an adhesive. The caudal glands are identical with the pedal glands of rotifers and like the latter secrete an adhesive cement by which the animal anchors itself to objects.

The cervical or ventral gland is generally regarded as of excretory nature and hence will be considered with that system.

Directly internal to the epidermis is the layer of body-wall musculature, consisting exclusively of longitudinal chords into longitudinal strips, four, eight, or two in number, depending on the number of chords; but as the chords tend to die out posteriorly so the number of muscle strips may vary in different body regions of the same worm. Very great weight was placed by A. Schneider (1860, 1866) on the form, number, and arrangement of the muscle cells as a taxonomic character and he coined the following terminology still in use. In some nematodes, there seem to be no definite muscle cells but the muscle layer is continuous or at best divided into two zones by the lateral chords; to this type Schneider applied the term *holomyarian*. It shows transitions to the formation of definite muscle cells, fusiform or rhomboid in shape, with their long axes parallel to the longitudinal body axis. When there are only a few longitudinal rows of muscle cells, mostly two to five, in each longitudinal strip, as in *Oxyuris* (Martini, 1916), the condition is termed *meromyarian* and when a larger number of rows is present in each strip, the term *polymyarian* is applied. As transitions occur between these various conditions and as the number of muscle rows diminishes toward the body ends, Schneider's distinctions are no longer regarded as of particular importance and are certainly not indicative of relationship between nematode genera. The number of muscle cells is constant in forms in which it has been investigated as *Oxyuris* which has eight rows of muscle cells with six to nine cells in each longitudinal row, making a total of 65 cells. Similar numbers have been found in related forms.

The histology of the muscle cells of nematodes, especially of *Ascaris,* has been the object of numerous investigations, most recently by Roskin (1925) and Mucller (1929). The cells is of elongated fusiform or rhomboid shape and is composed of two zones, a fibrillar zone and a protoplasmic zone containing the nucleus. The fibrillar zone, situated next the epidermis, consists of longitudinal ribbons or bands of homogeneous contractile substance alternating with noncontractile material containing supporting fibrils. The protoplasmic part also contains a network of supporting fibrils and often gives off processes toward nerve fibers or other structures. The muscle cells may be flat with the fibrillar region limited to a basal zone paralleling the epidermis, a condition termed *platymyarian;* or the protoplasmic zone may bulge into the pseudocoel with the fibrillar zone extending up its sides, a condition termed

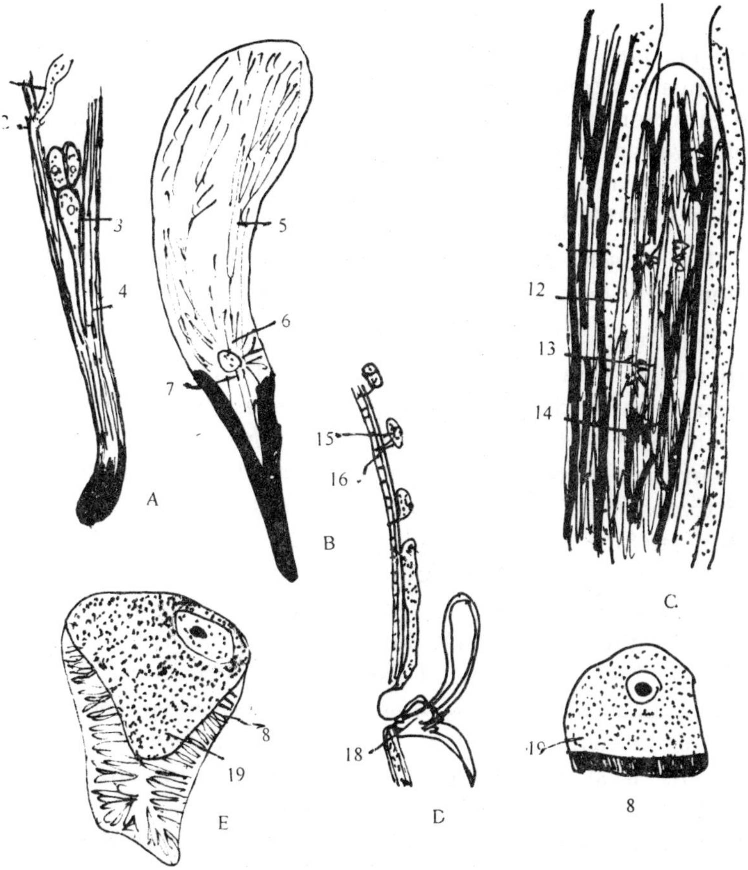

Fig. 1.14 : Epidermal glands, muscles. *A.* Caudal glands of a marine nematode (*Oncholaimus*) *B.* Details of a muscle cell (*Ascaris*). *C. Oxyuris equi* opened not along the ventral side to show arrangement of the muscle cells (meromyarian type). *D.* Preanal row of copulatory tubes with attached glands of male *Parasabatiera*. *E.* Coelomyarian type of muscle cell. *F.* Platymyarian type of muscle cell. 1, rectum; 2, anus; 3, caudal glands; 4, their duct; 5, supporting fibrils; 6, perinuclear basket of supporting fibrils; 7, nucleus; 8, muscle bands; 9, ventral chord; 10, muscle cells; 11, lateral chord; 12, posterior excretory canal; 13, dorsal chord; 14, large pseudocoelocytes (four); 15, gland cells; 16, their tubular exits; 17, spicules; 18, gubernaculum; 19, cytoplasmic part of muscle cell.

coelomyarian. It appears that the meromyarian and platymyarian conditions are the more primitive occurring in the simpler nematodes whereas the polymyarian and coelomyarian conditions are derived by development from the former, and characterize the larger parasitic nematodes such as the ascarids.

The muscles associated with the digestive and genital systems will be considered with those systems.

Pseudocoel.—The pseudocoel or space between the body wall and viscera is filled with fluid and usually contains fibrous tissue and fixed cells or nuclei, especially in parasitic nematodes. However, similar structures have been described in many free-living forms. Thus in *Thoracostoma* and *Enoplus* and *Cyclolaimus,* a network of nucleated strands covers the outer surface of the digestive tract and gonads. In *Oncholaimus* the pseudocoel is occupied by a fibrous nucleated network that forms membranes clothing the outer surface of the pharynx and the inner surface of the muscle layer. Distinct cells arranged more or less in rows may also occur, usually along one or more of the longitudinal chords. In parasitic nematodes there is often present dorsal to the pharynx a single large cell from which fenestrated membranes extend through the pseudocoel and over the viscera and muscle layer. Details of this condition for *Ascaris* have been given by Goldschmidt (1906) and for *Oxyuris* by Martini (1916). A similar dorsal cell and a nucleated fibrous network covering the digestive tract and inner surface of the body wall and sending strands to the longitudinal chords have been described for the oxyuroid nematode, *Cephalobellus papilliger,* by the Chitwoods (1933b). These authors further state (1937) that the pseudocoel cell or nucleus dorsal to the pharynx is common in parasitic nematodes and also in the free-living order Chromadoroidea, and that fibrous membranes in the pseudocoel are especially well developed in trichuroids and dioctophymatoids, notably in *Eustrongylides.* Common also in parasitic nematodes are two, four or six large cells occupying fixed constant positions, usually in relation to the longitudinal chords. These cells are often highly branched or stellate with rounded nodules along the sides and ends of the branches. In free-living

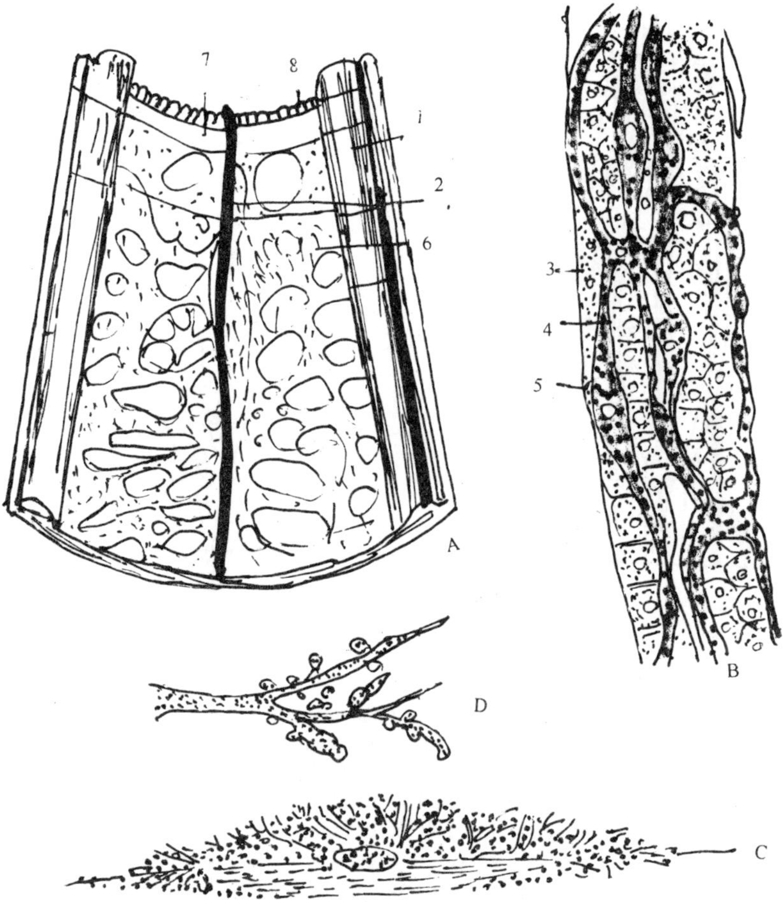

Fig. 1.15 : Membranes and cells of the pseudocoel. *A. Ascaris* opened out, showing fenestrated pseudocoel membrane. *B.* Net over the intestine of *Enoplus*. *C.* One of the four large pseudocoelocytes of *Contracaecum*. *D.* Branch of the same, showing nodules. 1, lateral chord; 2, ventral chord; 3, intestine; 4, network; 5, fatty granules; 6, fenestrated membrane; 7, nerve ring; 8, muscle cells.

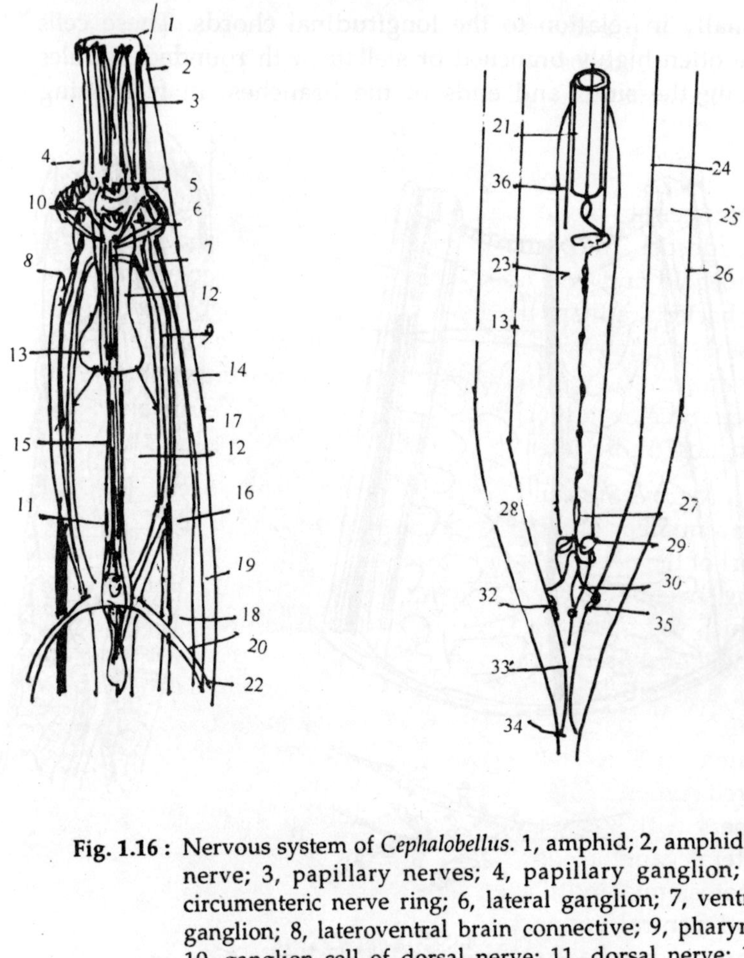

Fig. 1.16 : Nervous system of *Cephalobellus*. 1, amphid; 2, amphidial nerve; 3, papillary nerves; 4, papillary ganglion; 5, circumenteric nerve ring; 6, lateral ganglion; 7, ventral ganglion; 8, lateroventral brain connective; 9, pharynx; 10, ganglion cell of dorsal nerve; 11, dorsal nerve; 12, paired ventral nerves; 13, external lateral nerve; 14, internal lateral nerve; 15, ventrolateral nerves; 16, dorsolateral nerves; 17, anterior excretory canal; 18, transverse excretory canal; 19, excretory pore; 20, retrovesicular ganglion; 21, unpaired ventral nerve behind ganglion; 22, posterior excretory canal; 23, intestine; 24, prevulvar ganglion; 25, vulva; 26, postvulvar ganglia; 27, preanal ganglion; 28, intestinorectal sphincter; 29, rectal commissure; 30, dorsorectal nerve; 31, ventrolateral connectives; 32, anus; 33, caudal nerve; 34, phasmid; 35, lumbar ganglia; 36, vagina.

nematodes pseudocoel cells are generally smaller and more numerous. The *pseudocoelocytes,* as they may be called, and the fibrous strands and membranes associated with them may be regarded as a kind of mesenchyme.

The pseudocoelocytes are fixed and constant in position; there seem to be no free or wandering cells in the nematode pseudocoel. The pseudocoelocytes are neither phagocytic nor amoeboid and will not take up vital dyes. Injected particles or bacteria adhere to them but are not phagocytized. The visible particles in them appear to be chiefly of the nature of enzymatic bodies. According to Hurlaux, the giant pseudocoelocytes of parasitic nematodes have an oxidative function (more later).

Nervous System.—The nervous system has been described for a number of nematodes, chiefly parasitic ones. The main part of the nervous system consists of a ring, the *circumenteric ring,* encircling the pharynx, and of associated ganglia, of which the main ones are the paired lateral ganglia and the single or paired ventral ganglia. Others that may be present are a small dorsal ganglion, a pair of subdorsal ganglia, a pair of postlateral ganglia behind the lateral ganglia, and a pair of postventral ganglia behind the ventral ganglia. The circumenteric ring is commissural, that is, consists chiefly of fibers with very few ganglion cells. It is evident that the lateral ganglia correspond to the cerebral ganglia of other Protostomia and the circumenteric ring represents the dorsal and ventral connections between the lateral ganglia. The various ganglia are close to the ring and are connected with it or each other by short commissures. From the ring there proceed forward six papillary nerves, two dorso-lateral, two lateral, and two ventrolateral, to the sense organs of the anterior end of which it will be recalled there are three circlets with a total of 16 members. The ganglion cells of the papillary nerves are situated along the course of the nerves, usually near their entrance into the ring. The dorso- and ventrolateral papillary nerves on reaching the anterior end fork into three branches one of which supplies the inner labial papilla (or bristle) and the other two supply the outer labial and cephalic bristles of that sector. The lateral papillary nerves fork into only two branches as there are only two cephalic sense organs in the lateral sector. Although primarily sensory, the papillary nerves also contain motor fibers for the muscles that operate the bristles. The terminal branches of the papillary nerves are present even when the sense organs they originally supplied are absent or rudimentary as in many parasitic nematodes. There is also in the anterior end a pair of amphidial nerves from the amphids; they connect with the lateral ganglia, not with the circumenteric ring, and each also bears a ganglion. The amphidial nerves are larger with larger ganglia in the free-living nematodes.

From the nerve ring and associated ganglia there proceed posteriorly the following main nerves: a mid-dorsal nerve, one to three pairs of lateral nerves, and a mid-ventral nerve. The dorsal nerve originates from the middorsal region of the nerve ring, usually presents at once a small swelling, the dorsal ganglion, and then proceeds posteriorly in the dorsal chord as a motor nerve devoid of or poor in ganglion cells. The lateral somatic nerves originate mostly from the lateral ganglia and are of sensory nature with ganglionic swellings in their course along the lateral chords. They supply a sensory branch to the cervical papillae when present and posteriorly enter a pair of large ganglion in the anal region, the lumbar ganglia. Dorsolateral nerves may occur between the dorsal and lateral nerves. The ventral nerve is the main body nerve and is in reality a ganglionated cord. It is usually paired at its origin from the ventral part of the nerve ring and the two nerves proceed posteriorly in the ventral chord to a point behind the excretory pore where they usually fuse to form a ganglion, the *retrovesicular* ganglion, behind which the single ventral nerve continues as a ganglionated cord to the anal region. Here it enters a single or paired *anal* or *preanal* ganglion from which a pair of connectives, the *anolumbar* or ventrolateral connectives extend to the lumbar ganglia. The ventral cord usually passes to the right of the vulva in females but in some nematodes remains paired to a point behind the vulva where fusion to one cord occurs in a ganglion called the *postvulvar* ganglion. In some nematodes as *Ascaris* there are numerous ventrolateral connectives between the ventral and lateral nerves. Ventrolateral (subventral) nerves may also occur. The tail is supplied by nerves from the anal and lumbar ganglia and those from the latter also innervate the phasmids. There may be pair of caudal ganglia in the tail. The innervation of the posterior end is more complicated in males than in females. The genital papillae are innervated from bipolar sensory nerve cells that form a longitudinal strand, the *bursal* nerve, on either side in or near the lateral chords. The axones of these cells enter the lateral nerve and eventually reach the ventral nerve by way of the single or multiple ventrolateral connectives. The innervation of the bursa in bursate males has been described only for *Ancylostoma* in which there are two pairs of anal ganglia, a pair of cloacal ganglia, two pairs of ventrolateral connectives, and a chain of three lumbar ganglia; nerves from all these elements enter the bursa in which they branch.

The foregoing account is of general applicability to the various species for which the nervous system has been described but naturally the details differ in different species. The innervation of the anterior end is very similar in all but differences occur in the number of nerves behind the nerve ring. The greatest divergences from the usual plan are seen in *Ascaris* which is also the form in which the nervous system has been most thoroughly studied. *Ascaris* is peculiar in having a pair of subdorsal ganglia in addition to the single unpaired dorsal ganglion, in the subdivision of the lateral ganglia into six ganglia one of which is the ganglion of the amphidial nerve, in the large size of the paired ventral ganglia, and in the occurrence of numerous asymmetrical dorsoventral loops connecting the dorsal and ventral nerves throughout the body length.

In addition to the body nervous system there are in nematodes two enteric (or sympathetic) systems, one in the

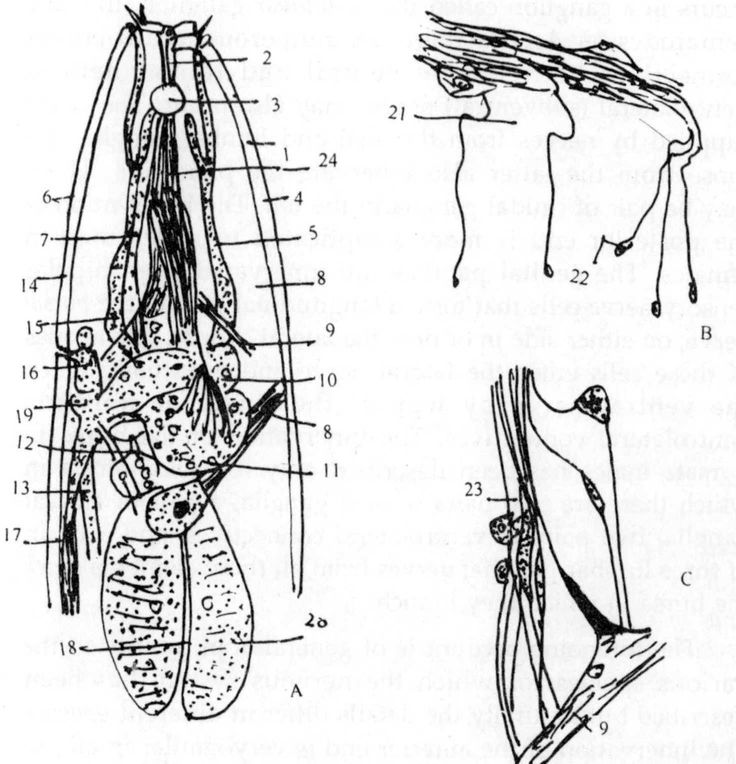

Fig. 1.17 : Nervous system (continued). *A.* Brain and anterior nerves of *Siphonolaimus*, lateral view. *B.* Bursal nerve with bipolar neurones supplying genital papillae. C. Papillary ganglion of an ascarid. 1, three branches of papillary nerve (only three of the six papillary nerves shown); 2, buccal capsule; 3, amphid; 4, amphidial nerve; 5, papillary ganglion; 6, amphidial gland; 7, pharynx; 8, pharyngo-cutaneous muscles; 9, circumenteric nerve ring; 10, lateral ganglion; 11, postlateral ganglion; 12, ventral ganglion; 13, postventral ganglion; 14, gland attached to bristle; 15, excretory pore; 16, ampulla of renette; 17, renette; 18, ventral nerve; 19, ventrolateral brain connective; 20, end bulb of pharynx; 21, bipolar cells; 22, genital papilla; 23, ganglion cells of papillary ganglion; 24, odontostyle.

pharynx and the other in the rectum. The pharynx system, connected directly with the ventral part of the circumenteric nerve ring, consists in general of three longitudinal strands, one in each sector of the pharynx wall. These strands contain nerve cells and are connected by two or three ring connectives. The rectal system in its simplest form consists of a pair of anorectal commissures extending from the anal ganglia to the dorsal surface of the rectum where they join a dorsorectal ganglion from which a median caudal nerve proceeds into the tail. In males the system is more complicated chiefly throughout the interpolation of a ganglion on each anorectal commissure from which ganglia nerves proceed to the cloaca. In *Ancylostoma* males, which have two pairs of anal ganglia, one behind the other, there are also two pairs of anorectal commissures and two dorsorectal ganglia in tandem arrangement.

An incredible amount of detail concerning the finer structure of the nervous system of *Ascaris* has bee furnished by Goldschmidt. According to this and similar work, the nerve cells are constant in number, location, shape, and course of their fibers. Each ganglion contains a definite fixed number of cells, e.g., there are seven bipolar sensory cells in each of the papillary ganglia of *Ascaris*. The sensory neurones of nematodes are bipolar, with one branch (dendrite) coming from the sense organ and the other (axone) proceeding along a nerve trunk.

Sense Organs.—As already noted, the nematodes, especially the free-living ones, are abundantly supplied with sense organs, chiefly in the form of bristles and papillae. Although the number and arrangement of the head sense organs has now been determined and depicted for a large number of nematodes by the study of *en face* preparations, very little exact information is available as to the histological construction of these organs. The labial and cephalic bristles are cuticular structures that are jointed to the cuticle. Each contains a nerve fiber that passes through an aperture in the cuticle to join the adjacent papillary nerve. Bristles may be supplied by a gland cell. The only satisfactory account of

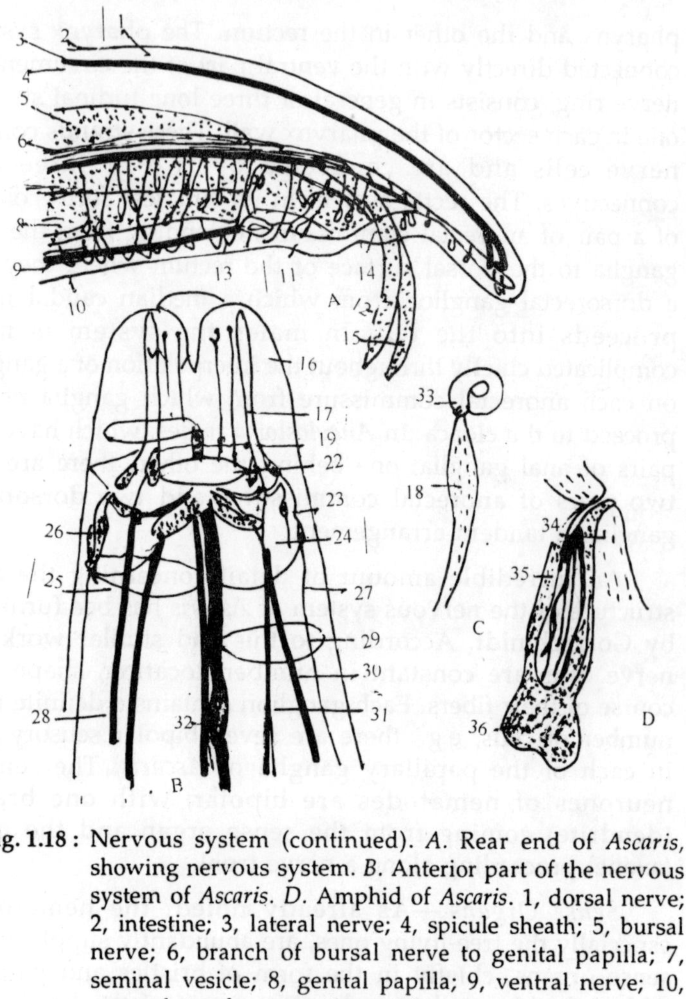

Fig. 1.18 : Nervous system (continued). *A.* Rear end of *Ascaris*, showing nervous system. *B.* Anterior part of the nervous system of *Ascaris*. *D.* Amphid of *Ascaris*. 1, dorsal nerve; 2, intestine; 3, lateral nerve; 4, spicule sheath; 5, bursal nerve; 6, branch of bursal nerve to genital papilla; 7, seminal vesicle; 8, genital papilla; 9, ventral nerve; 10, ventrolateral connectives; 11, preanal ganglion; 12, anorectal connectives; 13, anus; 14, spicules; 15, caudal nerves; 16, sensory papillae; 17, papillary nerves; 18, amphidial nerves; 19, papillary ganglion; 20, amphidial ganglion; 21, nerve ring; 22, dorsal ganglion; 23, subdorsal ganglion; 24, parts of the lateral ganglia; 25, ventral ganglia; 26, cervical papillae; 27, dorsal nerve; 28, ventral nerve; 29, dorsolateral nerves; 30, ventrolateral nerves; 31, dorsoventral connectives; 32, ganglion of ventral nerve; 33, amphid; 34, cuticle; 35, nerve endings; 36, supporting cell.

the structure of the labial and cephalic papillae is that of Goldschmidt (1903) for *Ascaris*. *Ascaris* has three lips formed by the fusion of the original 6 and each lip bears two papillae, each of which consists of 2 sense organs, making a total of 12. It appears that in *Ascaris* the 6 inner labial papillae are wanting and therefore the 12 sense organs represent the 6 outer labial papillae, the 4 cephalic papillae, and the 2 amphids (formerly erroneously called lateral papillae in parasitic nematodes). The cuticle over each pair of sense organs is greatly thinned and marked off by a circular groove. The medial sense organ of each pair (i.e., the outer labial papilla) consists of a stout nerve fiber that ascends toward the surface, narrows, and terminates beneath the cuticle in a bulb-like ending. The outer papilla of the dorsolateral and ventrolateral pairs (i.e., the cephalic papilla) is similar but just beneath the cuticle has a lens-shaped expansion from which a canal, at first narrow, then wider, passes to the surface. A nerve fiber runs along the center of this canal. In both kinds a papillae there is, shortly before they reach the under surface of the cuticle, a region that stains intensely, and in both the ascending nerve fiber is accompanied by two cells, an inner supporting cell that embraces the fiber and an outer accompanying cell that lies alongside the supporting cell. Despite these names, invented by Goldschmidt, the purpose of these cells is not clear nor can it be stated to what extend these descriptions for *Ascaris* apply to other nematodes.

The cervical papilla of *Ascaris* is similar to the outer labial papilla of the same animal, consisting of an ascending nerve fiber terminating in a little elevation beneath the cuticle and also provided with a supporting and an accompanying cell. The genital papillae (*Ascaris*) are wart-like elevations of greatly thinned cuticle with a central circular opening. The underlying sense organ consists of one to three ascending nerve fibers embedded in a supporting cell and terminating distally in a narrow canal that perforates the cuticle to open in the center of the papilla. It is supposed that all the types of papillae of nematodes are of a tactile nature.

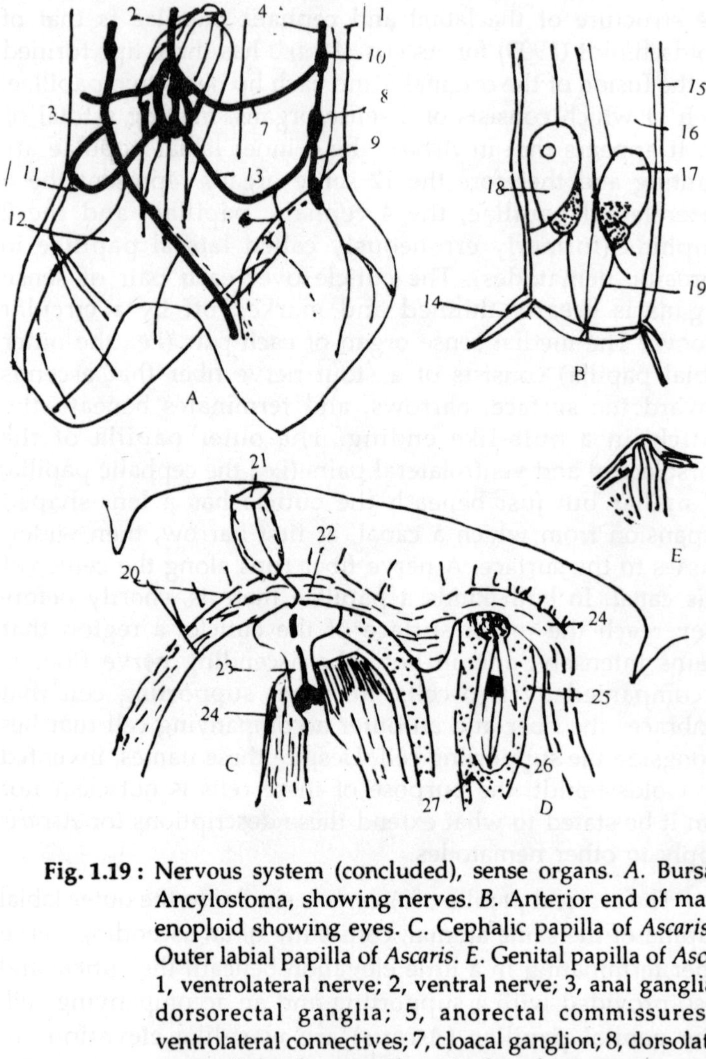

Fig. 1.19 : Nervous system (concluded), sense organs. *A*. Bursa of Ancylostoma, showing nerves. *B*. Anterior end of marine enoploid showing eyes. C. Cephalic papilla of *Ascaris*. *D*. Outer labial papilla of *Ascaris*. *E*. Genital papilla of *Ascaris*. 1, ventrolateral nerve; 2, ventral nerve; 3, anal ganglia; 4, dorsorectal ganglia; 5, anorectal commissures; 6, ventrolateral connectives; 7, cloacal ganglion; 8, dorsolateral nerve; 9, lumbar ganglia; 10, dorsal nerve; 11, median caudal nerve; 12, cloacal nerve; 13, rectal nerves; 14, nerves to bursal rays; 15, pharynx; 16, pit leading to eye; 17, lens-like body; 18, pigment; 19, retinal cell; 20, cuticle; 21, canal; 22, nerve fiber; 23, lens-like thickening; 24, pigmented region; 25, papillary nerve; 26, supporting cell; 27, accompanying cell.

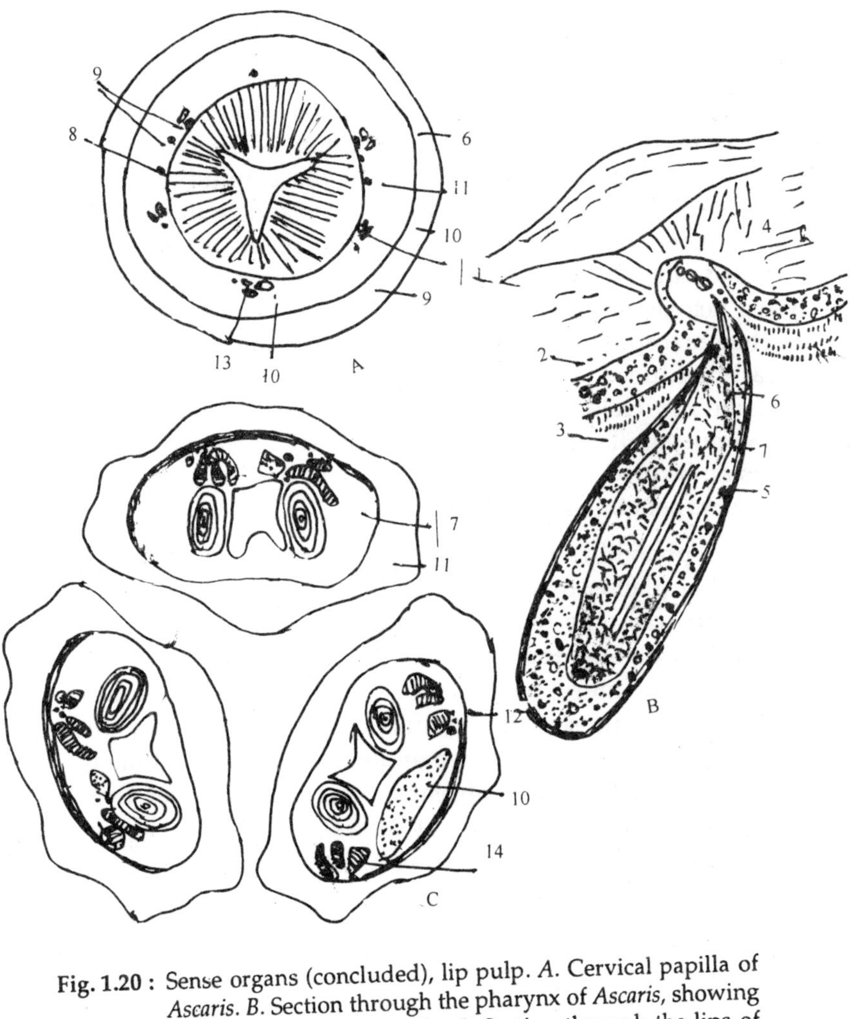

Fig. 1.20 : Sense organs (concluded), lip pulp. *A*. Cervical papilla of *Ascaris*. *B*. Section through the pharynx of *Ascaris*, showing rear ends of the pulp cells. *C*. Section through the lips of *Ascaris*, showing cell arrangement in the lip pulp. 1, cuticular cortex; 2, matrix; 3, fiber layers of cuticle; 4, sensory ending; 5, nerve; 6, supporting cell; 7, accompanying cell; 8, pharynx; 9, arcade cells; 10, fiber cells; 11, clavate cells; 12, sensory nerves; 13, filler cells; 14, cavity.

The amphids or lateral organs, as they were formerly called, occur in a lateral position outside the circlet of cephalic sense organs in free-living nematodes, usually incorporated into this circle tin parasitic members and hence erroneously mistaken for papillae. They are, as already noted, cuticular excavations of various shape; in the bottom of the excavation occurs a cluster or nerve endings of the amphidial nerve and also the opening of the amphidial gland, a unicellular gland often of great length that proceeds posteriorly alongside the pharynx. This description applies only to free-living nematodes in which the amphids are believed to serve as chemoreceptors. In parasitic nematodes, the amphids are greatly reduced and have lost the cuticular cavity. As described by Goldschmidt (1903) for *Ascaris* under the name of lateral papilla, the amphid consists of bundle of nerve fibers embedded in a supporting cell and uniting to a single fiber that penetrates the cuticle.

The phasmids when fully developed are a pair of unicellular glands that open by a canal and pore on either side of the tail. The two glands are called the precaudal or phasmidial glands. From this condition various degrees of reduction to mere papillae occur. The phasmids are presumably some sort of glandulo-sensory organ. They are best developed in parasitic nematodes but it is not definitely established that they are absent in marine and other free-living nematodes. It does appear certain that they are absent whenever caudal glands are present, and vice versa.

A pair of eyes is present in a few fresh-water and marine nematodes. The eyes are located on the side of the pharynx and usually consist of a lens-like cuticular body resting on a pigment cup. Little is known of the nervous part of the eyes. According to Schulz (1931) the lens and pigment mass in *Parasymplocostoma* rest upon a large cell of the nature of a sensory cell which probably has secreted the pigment. The eyes in this nematode are remarkable in being at the bottom of pits. The resemblance of this eye to the cerebral eyes of rotifers is striking. Schulz has also pointed out that in many nematodes what have been regarded as eyes are merely coloured patches or areas of excretory nature in the pharynx walls.

Nematodes are also abundantly supplied with free nerve endings, presumably of sensory nature. Some of these of regular occurrence in the anterior end of parasitic nematodes probably represent vanished sense organs of the head circlets.

Digestive System.—The mouth is terminal with rare exceptions and as already noted is bordered by the lips, six or fewer in number, sometimes altogether wanting, sometimes augmented by interlabia. Investigations of the interior tissue of the lips or *lip pulp* of ascarids have shown that this is composed of the distal ends of a definite number of unusually large, elongated cells, the main parts of which lie along the pharynx. In addition to the supporting and accompanying cells of the sensory fibers of the head papillae, the cell group in question comprises *clavate* cells, *fiber* cells, *ărcade* cells, and *filling* cells. The clavate cells, two in each lip, are long cells with a hollow interior, lamellate structure, and bulbous distal ends that fill the convexity of the lips. The fiber cells, so-called from their fibrous cytoplasm, are two in number in the dorsal lip (but fused distally) and single in the other lips. The arcade cells, nine in number, run along the pharynx, two opposite each angle, and one opposite each sector, and at the base of the lips are united by arches, whence the name arcade cells. From these arches processes extend into the lips but do not reach the main convexity of the latter which contains only the distal ends of the clavate cells. A few other cells are present as space fillers. In the anterior end of ascarids between the pharynx and the body wall, there are a total of 42 cells, the distal ends of most of which extend into the lips. According to the studies of Hoeppli (1925), the cell number and arrangement here is constant or practically so for a considerable number of parasitic nematodes but no comparable investigations on conditions in free-living nematodes are available.

The mouth leads into the *buccal capsule,* very variable in size, shape, and degree of differentiation in different nematodes. When well developed it is a cylindrical or prismatic tube, lined by cuticle often heavily sclerotized into ridges, rods, plates, and other stiffenings that often bear teeth of various sizes. In some genera as *Enoplus* there are three especially large and heavy teeth known as *jaws* or *gnathi*. The buccal capsule and its stiffenings often exhibit a triradiate arrangement corresponding with that of the pharynx. In some nematodes, especially rhabditoids, the buccal capsule is divisible into three sections (Steiner, 1933a): an anterior chamber enclosed by the lips, the *vestibule* or *cheilostom,* a middle and longest and most sclerotized portion, the *protostom,* and a small terminal chamber, the *telostom,* called *glottoid apparatus* in some nematodes. In some forms, the protostom is further subdivisible. For the sclerotizations the general term *rhabdions* was coined by Steiner. Whether this analysis of the structure of the buccal capsule, based on rhabditoids, is applicable to other groups of nematodes is doubtful. It follows the erroneous Chitwood theory that the rhabditoids are the most primitive existing nematodes whereas unquestionably the free-living marine nematodes occupy this position. In many marine nematodes the buccal capsule is small and weakly developed and when heavy sclerotizations occur, these seem to follow other structural plans than that of rhabditoids. Not infrequently the buccal capsule is wholly or partly embraced by pharyngeal tissue, being then spoken of as *embedded.*

In certain groups of nematodes, the buccal capsule is armed with a conspicuous protrusible *spear* or *stylet,* used to puncture plants and animal prey. This may be formed, as in tylenchoids, by the coming together of the sclerotizations of the buccal capsule, so that it constitutes a *buccal stylet.* This is necessarily hollow and forms the path of food intake. In other cases, as in dorylaimoids, the spear represents an enlarged tooth, that originates in the pharynx wall, may or may not be hollow, and may or may not lie in the path of food ingestion. This type will here be called *odontostyle.*

The pharynx (usually called esophagus by nematodologists) is one of the most characteristic features of nematode anatomy. It is a cylindrical tube of considerable length lined by a thin cuticle and bounded externally from the pseudocoel by a membrane; as noted above, the amphidial glands and the elongated cells that accompany the sensory nerves and contribute to the lip pulp embrace the anterior part of the pharynx. The pharynx lumen is triradiate, being extended into three symmetrically arranged longitudinal grooves that partially divide the pharynx wall into three sectors, one dorsal and two ventrolateral. The pharynx thus exhibits a marked triradiate structure continuous with that of the buccal capsule, lips, and head sense organs. The pharynx wall consists of a syncytial epithelium, which, however, is so thickly permeated with muscle fibers and gland tissue as to be scarcely demonstrable. Numerous radial fibers, usually arranged in clusters, extend from the lining cuticle to the bounding membrane. From the outer ends of the three grooves fibers believed to be of elastic nature ("marginal" fibers) proceed to the bounding membrane). Three highly branched uninucleate or multinucleate pharyngeal glands are typically present in the pharynx wall, one dorsal and two ventrolateral. They are so long and so greatly branched that portions of them are met with in almost any section through the pharynx. The main cuticularized duct of each gland opens, often by way of an ampulla, into the pharynx lumen, or may run forward into the buccal capsule to discharge there in connection with conspicuous structures as jaws or large teeth. It is usual for the dorsal gland to open much farther forward than do the ventrolateral glands. In the Enoploidea there are five or more pharyngeal glands instead of the usual three. In some tylenchoids, the glands protrude from the pharyngeal wall into the pseudocoel. Three small protruding *cardiac* glands occur at the rear end of the pharynx in some free-living nematodes as *Trilobus, Tripyla,* and related genera.

The pharynx frequently presents one or more muscular swellings known as *bulbs*. These are considered true bulbs

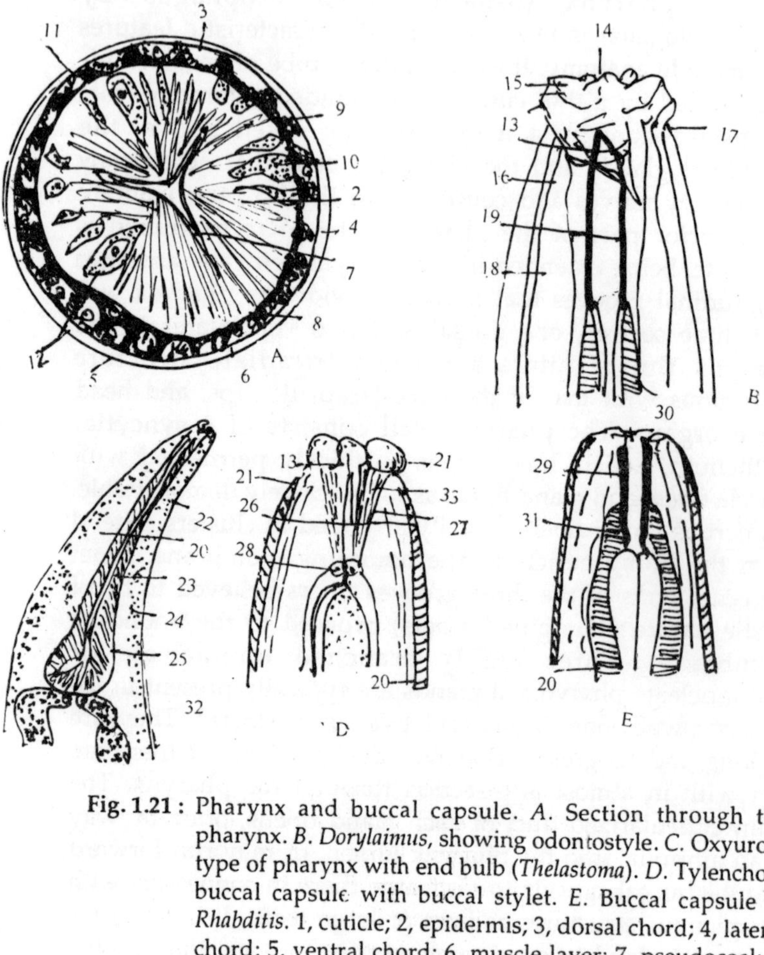

Fig. 1.21 : Pharynx and buccal capsule. *A.* Section through the pharynx. *B. Dorylaimus,* showing odontostyle. C. Oxyuroid type of pharynx with end bulb (*Thelastoma*). *D.* Tylenchoid buccal capsule with buccal stylet. *E.* Buccal capsule of *Rhabditis*. 1, cuticle; 2, epidermis; 3, dorsal chord; 4, lateral chord; 5, ventral chord; 6, muscle layer; 7, pseudocoel; 8, pharynx; 9, triradiate lumen; 10, muscle fibers of pharynx; 11, sections of pharyngeal glands; 12, marginal fibers; 13, lips; 14, labial papillae; 15, cephalic papillae; 16, amphid; 17, amphidial nerve; 18, odontostyle; 19, guide ring of odontostyle; 20, pharynx; 21, buccal capsule; 22, nerve ring; 23, excretory pore; 24, end bulb of pharynx; 25, intestine; 26, buccal stylet; 27, end knobs of stylet, derived from telostom; 28, duct of dorsal pharyngeal gland; 29, protostom; 30, cheilostom; 31, telostom; 32, pharyngo-intestinal valve; 33, protractor muscles of buccal stylet.

only when they contain a valvular arrangement in the form of three more or less sclerotized swellings that can partially close the lumen; otherwise they are called *pseudobulbs*. Bulbs or pseudobulbs may be situated about the pharynx middle, then spoken of as *median*, or may occur at the rear end of the pharynx, then termed *posterior, cardiac,* or *end* bulbs. With regard to shape and the presence of bulbs, pharynges are classifiable as follows: *cylindrical,* when of the same diameter throughout; *dorylamoid,* when slender anteriorly and wider posteriorly; *bulboid* or *oxyuroid,* when provided with an end bulb; *rhabditoid,* with an anterior wide region (*corpus*), usually leading into a median pseudobulb, followed by a narrowed region (*isthmus*), succeeded by an end bulb; *diplogasteroid,* with an anterior muscular region terminating in a median bulb, succeeded by a posterior glandular region; *tylenchoid,* similar to the diplogasteroid, except for the more slender anterior region; *aphelenchoid,* similar to the tylenchoid, except that the pharyngeal glands protrude posteriorly from the pharynx; and *mermithoid,* with a very long tubular, nonmuscular pharynx.

The Trichuroidea were formerly believed to have a very peculiar pharynx, supposed to consist of an intracellular lumen in a longitudinal row of large cells, the *stichocytes,* collectively termed the *stichosome.* It has been shown, however, by Chitwood (1930, 1935) that the trichuroid pharynx, though slender, is of the usual muscular type and is only embedded in or contiguous with the stichosome, the cells of which are pharyngeal glands opening into the pharynx. A stichosome is also present in the Mermithoidea.

The pharynx may be muscular throughout or its posterior part may be devoid of muscle fibers, being then spoken of as a glandular or *ventricular* region. A pharynx muscular anteriorly and glandular posteriorly is characteristic of the orders Spiruroidea and Filarioidea. In ascaroids, especially anisakines, the posterior part of the pharynx often gives off one or more blind, usually solid, diverticula or appendages. This appendage or pharyngeal caecum is single and directed backward in *Raphidascaris*, *Contracaecum*, and *Goezia*, whereas in *Multicaecum*, *Polycaecum*, and *Typhlophorus*, all three of which are parasites of crocodilian reptiles, there are two

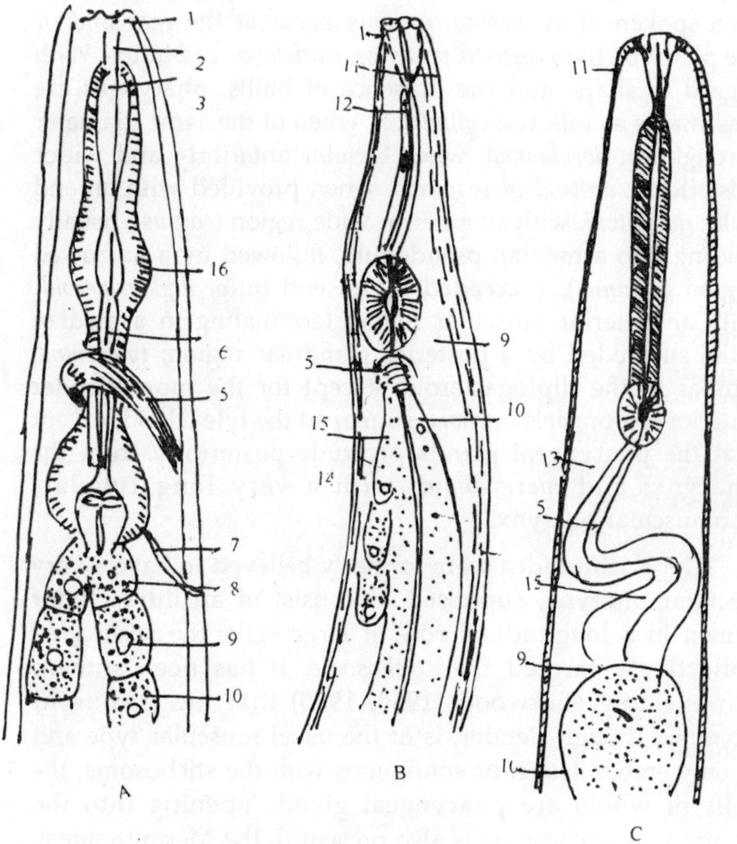

Fig. 1.22 : Pharynx (continued). *A*. Pharynx of *Rhabditis maupasi*, illustrating rhabditoid type. *B*. Aphelenchoid type of pharynx. C. Tylenchoid type of pharynx. 1, lips; 2, protostom; 3, telostom; 4, corpus of pharynx; 5, nerve ring; 6, isthmus of pharynx; 7, end bulb of pharynx; 8, valve apparatus of end bulb; 9, excretory pore; 10, intestine; 11, buccal stylet; 12, protractor muscles of same; 13, median bulb of pharynx; 14, pharyngeal glands; 15, glandular part of pharynx; 16, median pseudobulb.

anterior and three posterior pharyngeal caeca. No esophagus is present between the pharynx and the intestine but pharyngeal tissue usually projects into the beginning of the intestine as a pharyngo-intestinal valve.

The intestine or midgut, following directly upon the pharynx, is a simple straight tube formed of a one-layered cellular epithelium. In *Contracaecum* and several other anisakine ascaroids, its beginning gives off a blind hollow caecum that extends anteriorly alongside the pharynx. The intestinal epithelium consists of large cuboidal or columnar cells packed with inclusions and provided with a conspicuous rod border that probably represents altered immobilized cilia. The number of intestinal cells is limited and in many of the smaller parasitic nematodes only two rows are present. In some strongyloids, the intestinal cells are multinucleate. The rear part of the intestine may be covered externally with a delicate muscle net similar to that of rotifers. At its junction with the rectum the intestine is provided with a uninucleate sphincter termed the intestinorectal valve. In the Mermithidae, the intestine has altered to a blind, food-storing cylinder, that lacks a lumen, is not connected with either pharynx or anus, and consists of two or more rows of large cells filled with food inclusions.

The rectum or hindgut is a short usually flattened tube provided with rectal glands in many parasitic nematodes; these are large unicellular glands, usually three in number in females, six in males. Being of proctodaeal origin, the rectum is lined by cuticle outside which is an epithelium of a few large cells covered externally by muscle tissue. The anus is a transverse slit operated by a *depressor ani* muscle, frequently H-shaped. This muscle acts to elevate the dorsal wall of the rectum and the posterior lip of the anal aperture thus assisting in defecation. Rectum and anus are lacking in the females of some nematodes, as the mermithids and the guineaworm, but are always present in males as they serve as the outlet of the male reproductive system. The sperm duct enters the ventral wall of the rectum in all nematodes

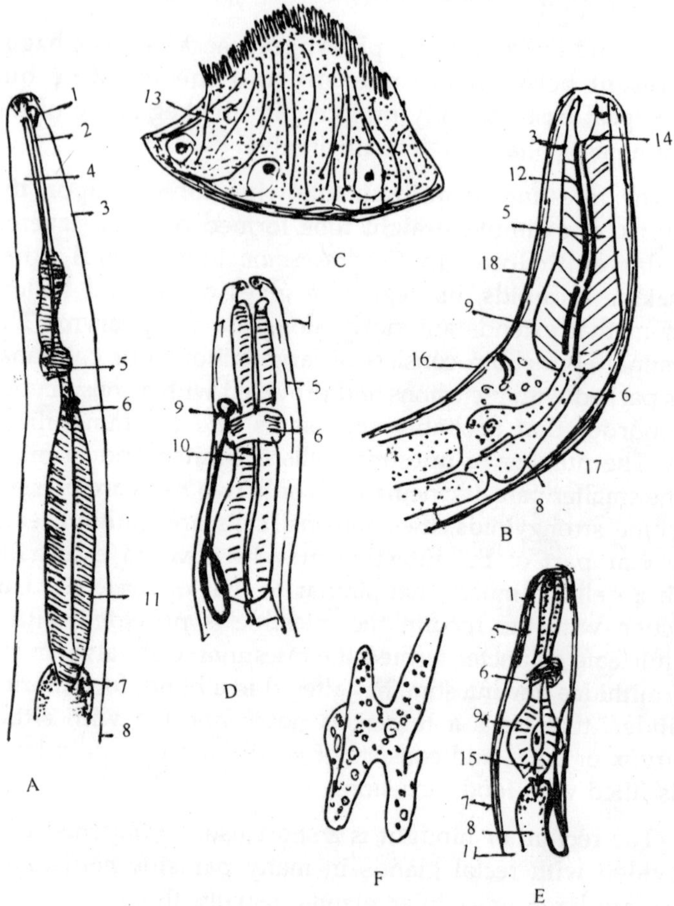

Fig. 1.23 : Digestive tract (concluded), excretory system. *A*. Dorylaimoid type of pharynx. *B*. Pharynx of *Diplogaster*. *C*. Section through nematode intestine. *D*. Anterior end of *Linhomeus*, showing renette with ampulla. *E*. Anterior end of *Desmolaimus*, with simple renette. *F*. Lobulate renette of *Enoplus*. 1, amphid; 2, odontostyle; 3, wall of buccal capsule; 4, ring guide of odontostyle; 5, pharynx; 6, nerve ring; 7, pharyngo-intestinal valve; 8, intestine; 9, excretory pore; 10, ampulla; 11, renette; 12, muscle of tooth; 13, rod border of intestinal epithelium; 14, tooth in buccal capsule; 15, end bulb of pharynx; 16, glandular part of pharynx; 17, cervical papilla; 18, median pseudobulb.

except the Trichuroidea in which is enters dorsolaterally. The rectum is thus in whole or in part a cloaca in male nematodes in which sex it is provided with a spicular pouch that contains the copulatory spicules.

Excretory System.—This system is peculiar in nematodes, being totally unlike that of other Aschelminthes or other Protostomia, for that matter. It is devoid of flame bulbs or any current-producing mechanism. The structures that are usually considered to be excretory (evidence later) are of two sorts, glandular and tubular. The glandular type of excretory system is found in free-living marine nematodes, which are here regarded as the most primitive existing members of the class. It here consists typically of a single large gland cell, situated ventrally in the region of the posterior end of the pharynx or anterior part of the midgut. This cell is usually called the *ventral* or *cervical* gland but Cobb's term *renette* has met with some acceptance. The long or short neck runs forward from the cell and opens in the mid-ventral line by an excretory pore. This pore is generally located in the region of the nerve ring but may be more anterior, even at the level of the lips. An ampullary enlargement may occur prior to the pore. The renette is lobulate in *Enoplus* and usually consists of a group of several cells in the Chromodoroidea. It is by no means present in all free-living marine nematodes. Little information is available as to its occurrence in fresh-water and terrestrial nematodes but those obviously related to marine forms are known to possess one. The Dorylaimoidea, a group of terrestrial spear-bearing nematodes, generally lack a renette although the excretory pore may be detectable. In the Mermithidae, a renette is present only in parasitic juvenile stages, not in adults, although these are free-living. A two-celled renette occurs in *Rhabdias* and many rhabditoids. From this condition appears but a step to the canal system by way of long canal-like outgrowths from the renette cells as in juvenile *Ancylostoma*. Later anterior outgrowths arise, and an H-shaped system results as in *Oesophagostomum*. The canals may pass through the renette cells as in

Oesophagostomum or the renette cells may depend from a transverse connection between the limbs of the H as in certain species of *Rhabditis*. In *Contracaecum* and other anisakine ascaroids there is but one renette cell and one canal (left), presumably through loss of the right part of the system; here the canal passes through the renette. Finally the renette cells become unrecognizable as such and H-shaped system of canals embedded in cytoplasmic material remains as in *Camallanus* and oxyuroids. When fully developed, the H-system thus consists of a longitudinal canal, the lateral excretory vessel, running in each lateral chord, a transverse canal connecting the lateral vessels by way of a bridge of a lateral chord tissue, and a common stem leading from the transverse vessels to the excretory pore. The posterior limbs or parts of the lateral canal posterior to the transverse canal, are usually much longer than the anterior limbs. The common stem may be enlarged to an ampulla or bladder-like reservoir. Through loss of the anterior limbs, the H-system becomes altered to the shape of an inverted U. In *Ascaris*, the H-system is present, but the anterior limbs are somewhat reduced and the transverse canal and adjacent parts of the lateral canals are altered into a network which is particularly evident on the left side where it contains the single nucleus of the system. A greater development of the left than of the right lateral vessel often obtains and as already noted only the left posterior limbs remains tin anisakine ascaroids. A loss of one lateral canal, either right or left, but usually left, is seen in the tylenchoids. The H-system of canals obtains in adult Strongyloidea and Spiruroidea or the inverted U-type through loss of the anterior limbs. The Trichuroidea and Dioctophymoidea are devoid of an excretory system in the adult stage but the inverted U-type is present in at least some Filarioidea.

On present evidence it is difficult to decide the relationship of the renette to the canal system. Stekhoven (1937-1939) believes the two structures are wholly independent but may sometimes secondarily establish connections. The Chitwoods consider the canal system to

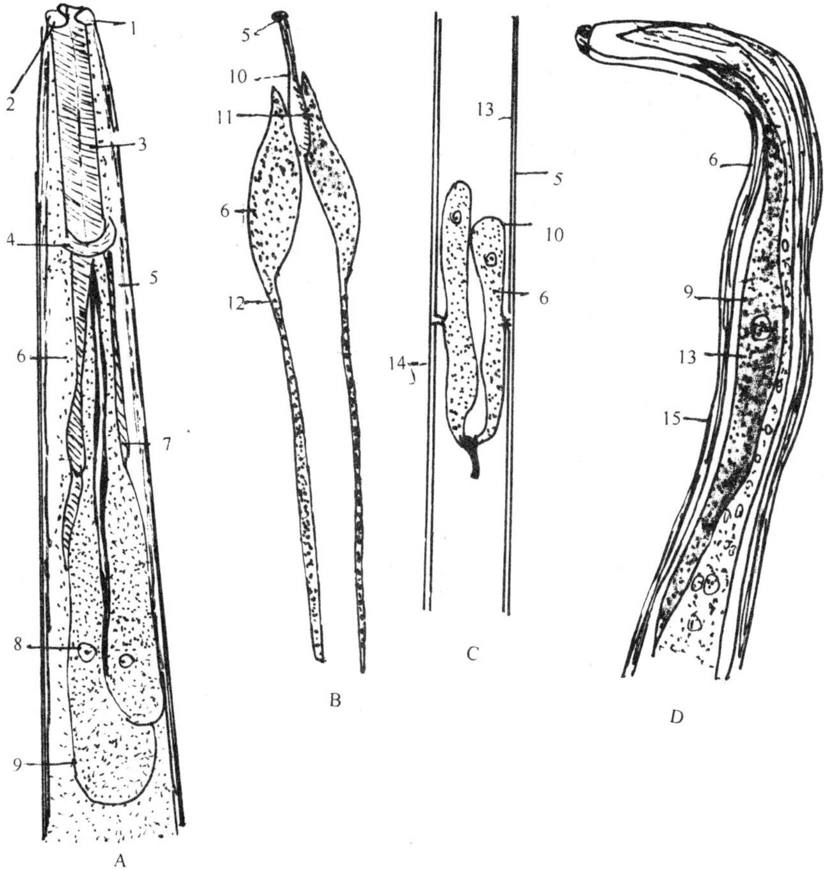

Fig. 1.24 : Excretory system. *A. Rhabdias* with two-celled renette. *B.* Juvenile *Ancylostoma,* with posterior canals developing from the renette cells. *C. Oesophagostomum,* with lateral connections of renette cells to canals. *D. Contracaecum,* with excretory system on one side only. 1, lips; 2, buccal capsule; 3, pharynx; 4, nerve ring; 5, excretory pore; 6, renette cells; 7, pharyngeal bulb; 8, intestine; 9, nucleus of renette cells; 10, common stem; 11, contractile ampulla; 12, posterior extensions of renette cells; 13, excretory canal; 14, connections between canals and renette cells; 15, pseudocoelocytes.

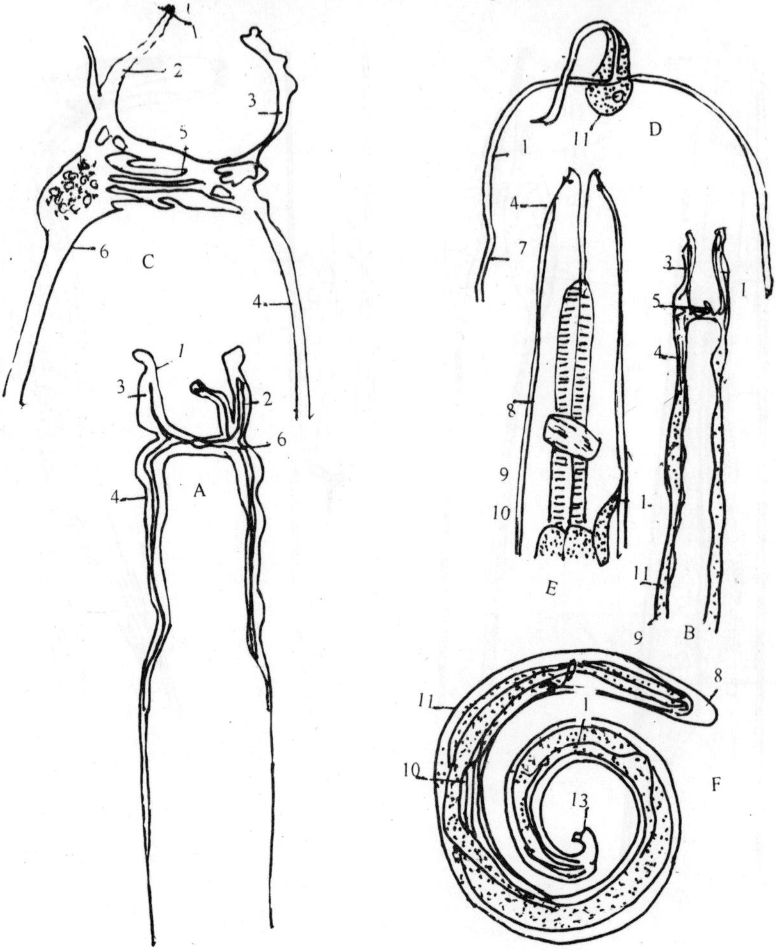

Fig. 1.25 : Excretory system (continued). *A.* H-type, *Camallanus. B.* H-type with many nuclei, *Cucullanus. C.* H-type, horse ascaris, with transverse canal altered into a network. *D. Physaloptera,* fourth stage juvenile, inverted U-type. *E.* Juvenile spiruroid (*Ascarops*) with one-celled renette. *F.* Juvenile strongyloid (*Metastrongylus*) with long one-celled renette. 1, excretory pore; 2, common canal; 3, anterior canal; 4, posterior canal; 5, transverse canal or network in C; 6, nucleus; 7, buccal capsule; 8, pharynx; 9, nerve ring; 10, intestine; 11, renette; 12, sheath; 13, anus.

be the original one from which the renette type can be derived by way of forms regarded as transitional that have both systems concomitantly. On such a view the lateral canals would presumably represent protonephridia that have lost their flame bulbs. This is an attractive theory that would reconcile the excretory system of nematodes with that of other Aschelminthes; but it meets the insuperable objection that the marine nematodes, which have the renette system, are undoubtedly the most primitive existing nematodes, ancestral to nematodes with a canal type of excretory system. That the renette is the original type of nematode excretory system is definitely proved by the fact that the juveniles of many parasitic nematodes, including oxyuroids, ascaroids, spiruroids, and strongyloids, have a typical one-celled renette, indistinguishable from that of marine nematodes (1890) followed the further history of this cell in *Enterobius* (an oxyuroid, with an H-type adult system) and found that the cell puts out two posterior processes (*B*), that undoubtedly become the posterior limbs of the lateral canals. This observation, plus the existence of transitional types that have both renette cells and canals, strongly indicates that the canals are intracellular outgrowths from the renette.

The vessels of the canal system are lined by a firm membrane outside of which is a layer of syncytial cytoplasm. Although sometimes containing numerous nuclei (*Cucullanus*), this cytoplasm is as a rule devoid of nuclei and the whole system usually bears a single nucleus, often very large, generally situated along the transverse canal, often on the left side. It thus appears to be a fact that in general the canal system represents channels in the cytoplasm of a single cell, presumably the altered renette. Additional nuclei along the common terminal duct are probably those of epidermal cells invaginated to form this duct. The view is therefore here adopted that the canals are intracellular outgrowths of the renette, probably necessitated by increasing body size, and that they are unrelated to protonephridia, which have disappeared in the evolution of nematodes.

Reproductive System.—The nematodes as a rule are dioecious, existing as separate males and females. Males are readily distinguished externally from females by their smaller size, curvature of the posterior end, and presence of bursae, genital papillae, and other accessory copulatory structures. In marine and most parasitic nematodes, the sexes occur in equal proportions but this is not the case with terrestrial and fresh-water members. Micoletsky (1921) examined 50 to 500 individuals of some 45 species of terrestrial and fresh-water nematodes and found a preponderance of females in nearly all cases. In 15 species no males at all occurred in the material and there were fewer than 50 males per 100 females in 75 per cent of the terrestrial species and in 50 per cent of the fresh-water ones. The scarcity or absence of males indicates a tendency towards hermaphroditism or parthenogenesis in nematodes from these types of habitats. Hermaphroditism is in fact not uncommon among terrestrial nematodes, especially rhabditoids, as shown by the studies of Maupas (1900), Potts (1910), and Honda (1925). The hermaphroditism is usually of the protandric type; the gonad first produces sperm that are stored and later fertilize the eggs subsequently developed by the same gonad. Hermaphroditism is usually called *syngony* in works on nematodes. Syngonic nematodes have the external appearance of females. It was shown by the above workers that genuine males occasionally appear in cultures of syngonic nematodes; in 13 species of rhabditoids, Maupas obtained 0.13 to 45 males per thousand females. Curiously enough, these males, although anatomically normal, have lost the sexual instinct and usually will not copulate; when they do so, the offspring contain many more males than is usual for the species. The sperm produced by syngonic females are usually insufficient to fertilize all the eggs matured by the ovary so that following a production of fertilized eggs there is a period of emission of sterile eggs. Sometimes the ovary again reverts to sperm production. In some hermaphroditic nematodes, however, the production of sperm continues throughout the egg-laying

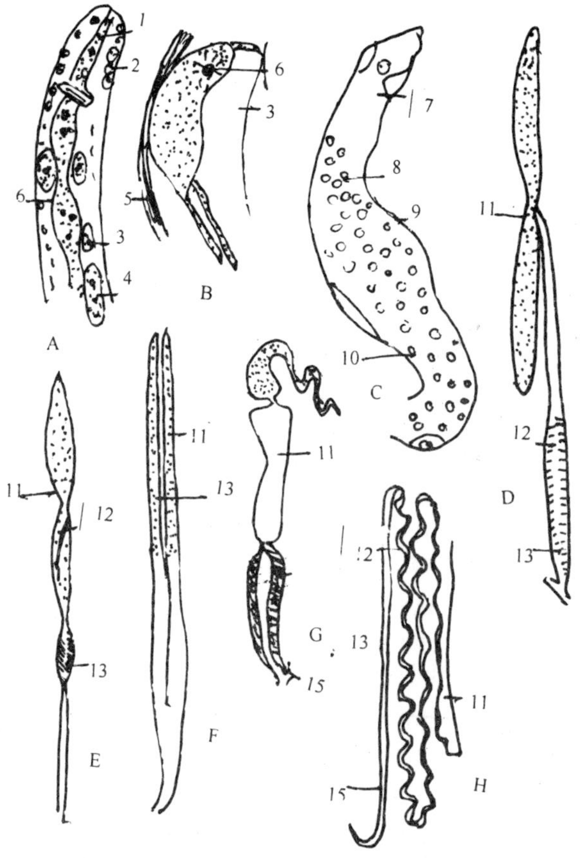

Fig. 1.26 : Excretory system (concluded), male reproductive system. *A. Enterobius* juvenile with one-celled renette. *B.* Same, later, with posterior canals growing from renette. *C.* Tip of gonad of pig ascaris. *D.* Scheme of male system with two opposite testes. *E.* Male system of *Antiocoma typica* with testes in tandem. *F.* Male system of *Heterodera marioni* with parallel testes. *G.* Male system with one anterior testis. *H.* Scheme of the male system of ascaris, with long coiled testis and duct. 1, pharynx; 2, nerve ring; 3, renette; 4, gonad primordium; 5, lateral chord; 6, excretory pore; 7, apical cell; 8, terminal cell; 9, germinal zone; 10, covering epithelium; 11, testis; 12, sperm duct; 13, seminal vesicle; 14, ejaculatory duct; 15, intestine. Spicules are omitted.

period, simultaneously with the production of eggs. In the experiments of Potts and Maupas, in which some syngonic species were bred for nearly 50 generations, there was no indication of deterioration as the result of such inbreeding.

The occurrence of parthenogenesis has been proved for several terrestrial nematodes, e.g., *Mermis subnigrescens* and *Heterodera marioni,* the root-knot nematode, although males are known in both species. Tyler bred 12 generations of females parthenogenetically in *H. marioni.* A remarkable kind of parthenogenesis has been recorded for *Rhabditis monhystera* by Belar (1924) and probably also exists in other species of *Rhabditis.* These species are dioecious and the eggs will not develop unless insemination by a male has occurred; but in those eggs that are to develop into females, the sperm merely acts as a stimulator to development and does not participate in the subsequent mitotic processes.

Intersexes, that is, females having some male characters, have been frequently reported in nematodes and seem to result from partial sex reversal under environmental influence.

Nematode gonads are of tubular shape, varying greatly in length, and may be straight, sinuous, reflexed, or coiled back and forth. As regards mode of germ-cell production, they are of two types, *telogonic* and *hologonic.* In hologonic gonads, limited to the orders Trichuroidea and Dioctophymoidea, proliferation of germ cells takes place along the entire length of the gonad, either throughout its walls or in limited areas; in telogonic gonads, germ cells are proliferated only in the blind proximal end, termed the *germinal zone,* usually from a single large *terminal* cell. The germinal zone is succeeded by a *growth zone* in which the gametogonia enlarge and differentiate. They are here not loose but attached to a central cytoplasmic strand or *rachis* of uncertain function. Further distally, the gametocytes become free from the rachis and undergo maturation processes. The gonad is covered by a flat epithelium continuous with the epithelium that forms the gonoduct wall. At the proximal tip of the gonad, this covering epithelium

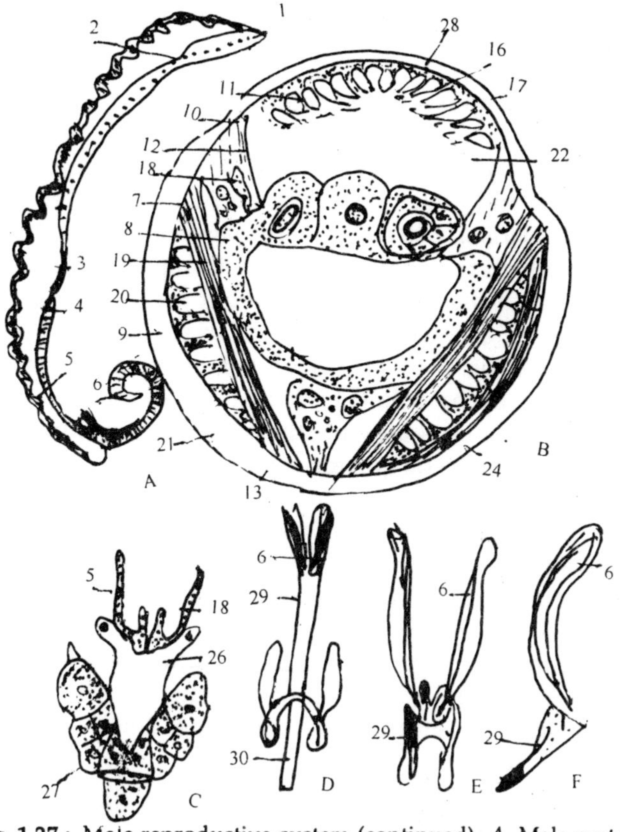

Fig. 1.27 : Male reproductive system (continued). *A.* Male system of *Trichuris. B.* Section through the rear end of ascaris. *C.* Development of spicule sacs. *D.* Male armature of a trichostronyloid, with talamon. *E.* Spicules and gubernaculum of a chromadoroid. *F.* Same from side view. 1, testis; 2, seminal vesicle; 3, ejaculatory duct; 4, cirrus; 5, intestine; 6, spicule; 7, cuticle; 8, epidermis; 9, muscle layer; 10, pseudocoel; 11, dorsal chord; 12, lateral chord; 13, ventral chord; 14, muscle layer of spicule pouch; 15, epidermis of spicule pouch; 16, spicule; 17, cytoplasmic core of spicule; 18, rectal glands; 19, cuticular lining of rectum; 20, lining epithelium of rectum; 21, copulatory muscle; 22, bursal nerve; 23, lateral nerve; 24, ventral nerve; 25, rectum; 26, cloaca; 27, evaginations to form spicule pouches; 28, dorsal nerve; 29, gubernaculum; 30, telamon.

may present a large *apical* cell, not to be confused with the true terminal proliferative cell.

Probably there were originally two gonads in all nematodes but in males a single testis is usually present and this extends anteriorly. However, two testes occur in many nematodes and these are oppositely oriented except in *Heterodera marioni* where they are parallel, and in *Anticoma typica*, in which they have a tandem arrangement. The terms *diorchic* and *monorchic* are convenient for referring to the two-testes or one-testis conditions, respectively. The free end of the testis is often convoluted or recurved and very long testes may be would back and forth. There is a single sperm duct continuous with the surface epithelium of the testis. This proceeds posteriorly and usually presents a widened, somewhat muscularized region of varying length, the *seminal vesicle*, posterior to which, the terminal, often heavily muscularized part of the sperm duct, called *ejaculatory duct*, continues to its junction with the rectum. The various divisions of the sperm duct may be marked off by constrictions and a sphincter may occur at the junction with the rectum. A varying number of prostatic glands, producing an adhesive secretion, may enter the distal end of the ejaculatory duct.

Male nematodes with very few exceptions (example, *Trichinella*) are provided with copulatory spicules secreted by and lodged in spicule pouches formed of cloacal evaginations that have the same histological structure as the cloacal wall. The spicule pouches, usually two in number, unite prior to their entrance into the cloaca. The spicules consist of hard sclerotized cuticle with a cytoplasmic core. They are typically and primitively two in number, but single in some nematodes as a result of fusion or, less often, through the loss of one spicule. The spicules are usually very short relative to the body length but are sometimes very elongated and exceptionally may equal or exceed the body length. They present a great variety of size and shape and may be styletiform, curved, flattened, provided with flanges, blade-like etc. The two spicules may be equal in size and alike in shape or may exhibit various degrees of dissimilarity. When unlike, the left spicule is usually long and slender, whereas the right one is short and stouter. Often the spicules are accompanied by an accessory piece, the *gubernaculum,* which is a sclerotization of the dorsal wall of the spicule pouch, very variable in shape but characterized in general by incurved margins. It acts to direct the spicules toward the anus and to prevent their piercing the cloacal wall. In Strongyloidea there is sometimes present another piece, the *telamon,* named and first described by Hall (1921); this is a slightly sclerotized, immovable formation of complicated shape in the ventral and lateral cloacal walls that also aids in directing the spicules during copulation. The morphological variations of the spicules and their accessory pieces furnish important taxonomic characters in nematodes, being utilized for generic and specific diagnoses in the same way as are the trophi of rotifers.

The spicules serve in copulation primarily to spread open the female gonopore and apparently do not act as conduits for the sperm. They are operated by retractor muscles extending from their proximal ends forward to the lateral chords and protractor or exsertor muscles origination on the ventral body wall. The gubernaculum is provided with similar muscles. Copulatory muscles in the form of a series of bands in the ventrolateral body wall in the anal region are of common occurrence in nematodes.

The order Trichuroidea is unique among nematodes in that the very long cloaca, often lined with spines, is eversible in males as a cirrus. A single copulatory spicule accompanies the cirrus in the Trichuroidae but is absent in other Trichuroidea.

The sperm of nematodes are usually unlike those of other animals, being rounded, conical, lobulate, or elongated bodies that move slowly in amoeboid fashion. Even when the external shape approaches the flagellate type, the cytological structure characteristic of flagellate sperm is wanting.

In females there are usually two ovaries oriented in opposite directions and these may be straight, sinuous, reflexed, or if very long wound back and forth. The ovaries and ducts may also be arranged in a parallel position extending either anteriorly or posteriorly, and reduction to a single ovary and duct is not uncommon. The terms *monodelphic* and *didelphic* are convenient for indicating the single and double condition of the female tract, respectively. The *polydelphic* condition, in which there are more than 2 (up to 10 or 11), also occurs, especially in the Physalopteridae (Spiruroidea).

The epithelial covering of the ovary continues distally as the oviduct which at first may be slender with a tall epithelium but which soon widens into a broad tube, the uterus, lined with a flat to cuboidal epithelium and provided with circular and oblique muscle fibers. The beginning of the uterus usually functions as a seminal receptacle in which sperm are stored and in which fertilization occurs; the seminal receptacle may also be otherwise located, often at or near the distal end of the female tract, sometimes as a lateral pouch. The seminal receptacle may have distinctive epithelium, as in *Ascaris,* where it is provided with tall processes that almost fill the lumen and apparently serve to phagocytize excess sperm. The uterus functions for the storage of eggs which, as they pass distally along it, undergo fertilization, shell formation, and some degree of embryonic development, in some cases as far as the formation of juveniles. Distally the uterus or uteri enter a common muscularized tube lined with cuticle, the vagina, usually quite short, that opens to the exterior by the female gonopore or vulva. The true vagina, recognized by its cuticular lining, may be extended inward by appropriating the distal part of the uterus which then comes to resemble the vagina histologically except for the absence of the cuticular lining; this vaginal extension is sometimes called *vagina uterina* or *uterine vagina*. In many strongyloids and spiruroids, the terminal part of the female tract becomes heavily muscularized, forming the *ovejector* (also spelled ovijector), functioning in the expulsion of the eggs. This may involve

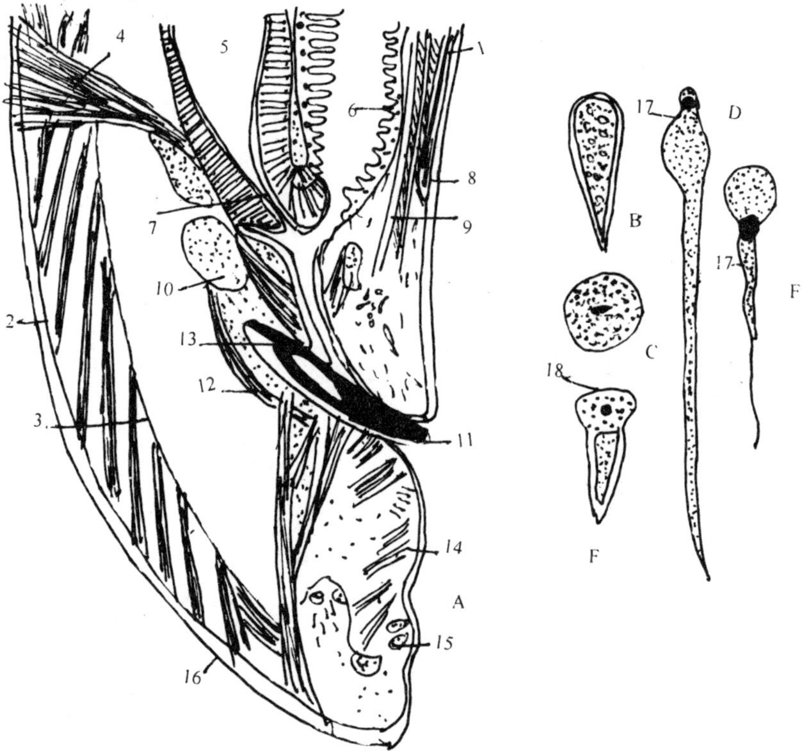

Fig. 1.28 : Male system (concluded). *A*. Section through the tail end of ascaris, showing relation of spicule to rectum. *B–F*. Types of nematode sperm. *B. Thoracostoma. C. Anaplostoma. D. Trilobus. E. Passalurus. F.* Horse ascaris. 1, cuticle; 2, copulatory muscles; 3, pseudocoel; 4, dilator of intestine; 5, intestine; 6, ejaculatory duct; 7, sphincter; 8, ventral nerve; 9, rectum; 10, rectal glands; 11, spicule; 12, spicule sheath; 13, protractor of spicule sheath (exsertor of spicule); 14, retractor muscles; 15, genital papillae; 16, nerve cells; 17, nucleus; 18, mitochondria.

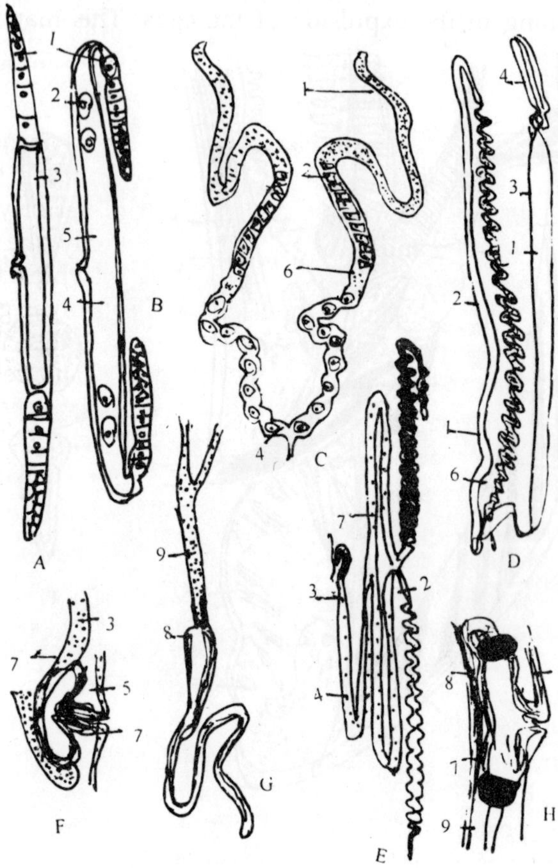

Fig. 1.29 : Female reproductive system. *A.* Scheme of female system with opposite straight ovaries. *B.* Scheme with opposite reflexed ovaries. *C. Heterodera,* with anterior parallel ovaries. *D. Trichuris,* with single sinuous ovary. *E. Heterakis gallinae* with opposite sinuous ovaries and simple ovejector. *F.* Ovejector of *Oesophagostomum. G.* Ovejector of *Tetrameres. H.* Ovejector of *Nematodirus.* 1, ovary; 2, oviduct; 3, uterus; 4, vagina; 5, vulva; 6, seminal receptacle; 7, ovejector; 8, sphincter; 9, non muscular proximal region

only the true vagina or also the uterine vagina and may be entirely single or partly forked. In its most complicated form, the ovejector consists of three regions, the terminal muscularized region, a sphincter, and a nonmuscular proximal region continuous with the uterus.

The female gonopore is a mid-ventral transverse slit with bulging lips that is usually situated along the middle third of the body but may be shifted anteriorly or to the vicinity of the anus. There is apparently only one nematode, *Rondonia rondoni,* a parasite of fishes, in which the vagina opens into the rectum. Both vaginal and vulvar glands may occur, chiefly in free-living marine nematodes. The vulvar glands consist of one to several pairs of gland cells situated anterior and posterior to the vulva. Dilator muscles in the form of radiating bands encircle the vulva in most nematodes.

A peculiar system of tubes, the *demanian* system, is found in the posterior part of many females of the marine family Oncholaimidae (order Enoploidea). This system, discovered by de Man in 1886, has since been investigated by Zur Strassen (1894), Cobb (1930) and Kreis (1934). It is variable in form and degree of development in different members of the family and reaches its highest complexity in *Metoncholaimus* and *Adoncholaimus.* In these, the system begins as a blind cell cluster in the dorsal wall of the midgut and runs posteriorly dorsal to the midgut as a thin-walled tube that gives off one or two connections into the adjacent uterus or uteri. These connections are encircled by a radiating cluster of gland cells termed the *rosette.* More posteriorly the tube forks and each fork, accompanied by a row of gland cells, the *moniliform* glands, opens to the exterior laterally shortly in front of the anus by a minute pore or set of pores. In *Adoncholaimus* in addition to two uterine connections, there is also an outlet into the vagina. In other oncholaims, the system is simpler with fewer parts and connections. According to Cobb, the system emits an adhesive fluid, presumably playing some role in reproduction.

The female tract is in general much longer and more complicated in parasitic than in free-living nematodes, in relation to the greater egg production characteristic of parasites. Especially the uteri and the growth zone of the ovary are greatly lengthened in parasitic nematodes.

Copulation and Reproductive Habits.—Although parthenogenetic and syngonic reproduction occur in nematodes, most species are dioecious and insemination by a male is necessary for the development of the eggs. In copulation, the spicules are inserted in the vulva and presumably the various accessory copulatory structures, as well as adhesive secretions, assist the process. The male is usually oriented at an angle to the female and often coils around her. In bursate males, the bursa folds over the female body. The trichuroid worm, *Trichosomoides crassicauda,* parasitic in the urinary tract of rats, is notable in the animal kingdom in that the minute males, 1.5 to 3.4 mm. long, dwell in the vagina of the female, 10 to 13 mm. long, where they were originally taken to be young worms.

The sperm travel up the female tract to the seminal receptacle where they fertilize the ripe ovocytes as these proceed distally. The entire sperm enters the egg. As soon as fertilization has occurred, a fertilization membrane appears around the egg and this thickens to form a shell. To the inner side of this shell, the egg soon gives rise to an inner or *lipoid* membrane, thin and delicate. As the eggs pass along the uterus there is added outside the shell a third or *protein* membrane, a product of the uterine wall. It is the protein membrane that furnishes the surface sculpturings, filaments, etc., present on some nematode eggs. The nematode egg is therefore typically surrounded by three membranes, the lipoid membrane, the shell, and the protein membrane; the last is not infrequently wanting. The chemical composition of these membranes has been repeatedly investigated. It was first shown by Krakow (1892) that the shell consists of chitin, something that is rather surprising in view of the complete lack of chitin elsewhere in the nematode body. The manner of formation of the shell and the lipoid membrane was elucidated by Fauré-Fremiet (1912) and Wottage (1937). The ripe ovocyte contains glycogen and lipoid globules and, immediately after fertilization, the glycogen globules move to the surface where they break

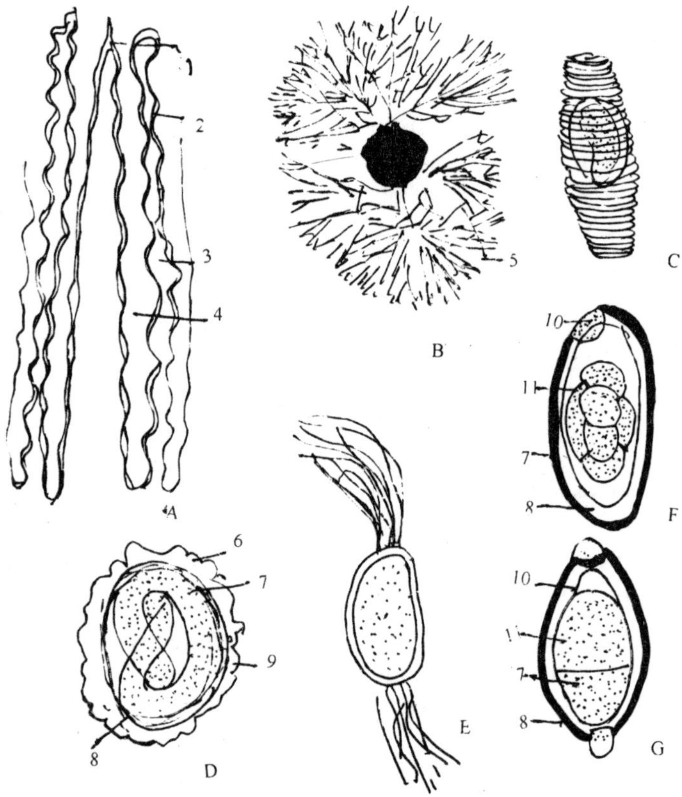

Fig. 1.30 : Female reproductive system, egg membranes. *A*. Scheme of the female system of ascaris. *B*. Mermithid egg. *C*. *Pseudonymus* egg enclosed in spiral filaments. *D*. Ascaris egg. *E*. Egg of *Tetrameres* with polar filaments. *F*. Oxyuroid egg (*Dermatoxys veligera*) with one opercular plug. *G*. Trichuroid egg (*Trichuris suis*) with two opercular plugs. 1, vagina; 2, uterus; 3, oviduct; 4, ovary; 5, byssus; 6, protein membrane; 7, shell; 8, lipoid membrane; 9, juvenile; 10, opercular plug; 11, embryo.

down into glucose from which glucosamine, the chief constituent of chitin, is synthesized. Next globules of saponified lipoid material move to the egg surface and become the lipoid membrane which when finished is composed of a sterol, cholesterol in at least some nematodes. Chitwood (1938) and Jacob and Jones (1939) verified these results and also proved the protein nature of the outermost membrane.

The shell enclosing the ripe egg is usually of rounded, oval, or elliptical shape with a smooth, spiny, or warty surface. A single filament occurs at each pole of the shell in *Citellina,* a bunch of filaments similarly located in *Tetrameres* and some other spiruroids, and in mermithoids there is at each pole an elaborately branched projection, the *byssus.* These projections of the protein membrane seem to function for attaching the eggs to objects. The eggs of *Pseudonymus,* an oxyuroid parasite of water beetles, are encircled by spirally coiled filaments that on contact with water unroll and become entangled in water plants. In some nematodes, provision is made for the easy escape of the young worm by a hiatus in the shell filled in with a soft opercular plug, single in oxyuroids present at both poles in trichroids and dioctophymoids.

Little seems to be known of the reproductive habits of free-living marine nematodes; apparently they lay a small number (less than 50) of relatively large eggs into their habitat, sometimes conglomerating them into masses by means of an adhesive secretion. Similar habits obtain in terrestrial nematodes, except that usually more eggs are laid, up to several hundreds. Phytoparasitic nematodes often lay their eggs in plant tissues or inside the plant galls that they evoke or accumulate them in their bodies which then dry up to form cysts. The zooparasitic nematodes are usually oviparous, discharging their eggs inside the host body. The eggs may be unsegmented on leaving the female or may undergo various degrees of development inside her uterus, up to the formation of juvenile worms. Development to young worms may also occur in the host body during the passage of the eggs to the exterior. A good many zooparasitic nematodes are ovoviviparous, that is, the eggs hatch to juvenile worms in the maternal uterus; the egg membranes are then more delicate than in oviparous forms and the protein membrane is usually absent. Whether true viviparity, the nourishing of embryos by the uterus, occurs in nematodes is doubtful.

It is usual for nematodologists to speak of eggs containing embryos or young worms as *embryonated.* While this term is convenient it is more exact to use the expression ovic embryos or juveniles.

Embryology.—The embryology of nematodes has been investigated by a number of zoologists and in fact has been the subject of several classical studies in cytology and germ-cell lineage. Although the cleavage does not follow the typical spiral plan, the development is nevertheless highly determinate in character and the fate of each cell can be definitely followed. There is further in nematodes a very early segregation of the germ-cell line, at the first cleavage, in fact. The principal embryological studies are those of Zur Strassen (1896), Boveri (1899), and Müller (1903) on *Parascaris equorum* (= *Ascaris megalocephala*), Neuhaus (1903) on *Rhabdias bufonis* (= *Rhabditis nigrovenosus*), Martini (1903, 1906) on *Camallanus lacustris* (= *Cucullanus elegans*), and Pai (1928) on *Turbatrix aceti* (= *Anguillula aceti*).

The egg is of oval form and in *Turbatrix* one pole bears the micropyle through which the sperm enters; this pole becomes the anterior end. The first cleavage occurs transverse to the egg axis and produces two cells of about equal size, an anterior cell called S_1 and a posterior cell P_1 which is the ancestor of the germ cells. At the second division S_1 divides into A and B, and P_1 divides into S_2 and P_2. In *Parascaris,* these four cells are at first arranged in a T-shape but later shift to the rhomboid shape characteristic of the four-cell stage of other nematodes. In this rhomboid, A is anterior, B, dorsal, S_2, ventral, and P_2, posterior. The S_1

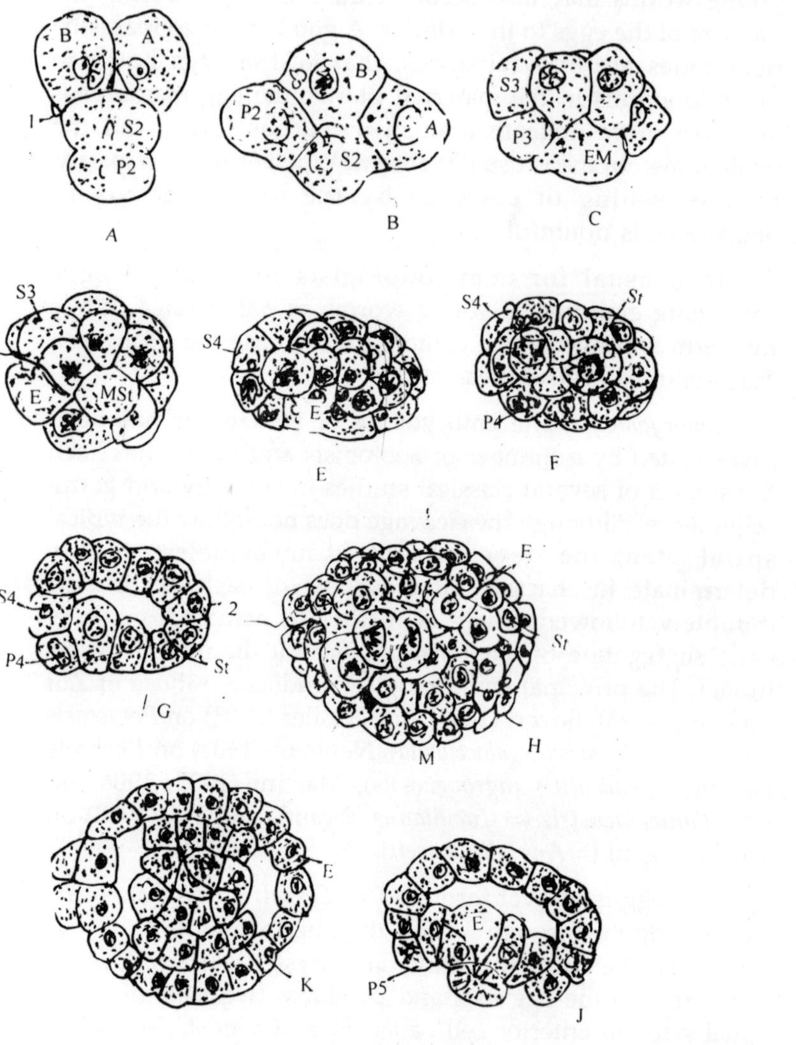

Fig. 1.31 : Embryology. *A*. T-shaped four-celled stage. *B*. Later four-cell stage, rhomboid shape. *C*. Seven-cell stage. *D*. 18-cell stage. *E*. Later stage, about 30 cells. *F*. Same as *E*, ventral view. *G*. Section through the blastula, sagittal. *H*. Gastrula, ventral view. *J*. Cross section through early gastrulation stage. *K*. Frontal section of the gastrula. 1, polar body; 2, blastocoel.

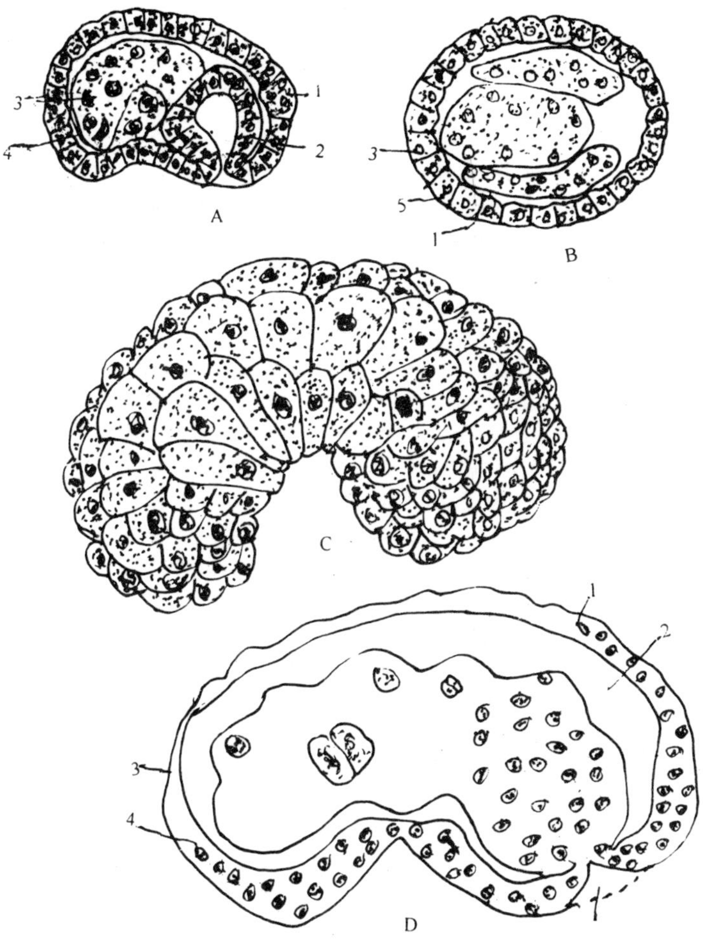

Fig. 1.32 : Embryology (continued). *A*. Sagittal section, showing formation of the stomodaeum. *B*. Frontal section of similar stage, showing mesodermal bands to either side of entoderm mass. *C*. Later stage of the embryo, worm shape now evident. *D*. Sagittal section of same, showing digestive tract indicated. 1, epidermis; 2, stomodaeum; 3, midgut entoderm; 4, germ cells; 5, mesoderm bands; 6, future mouth.

cell is the primary somatic cell and the descendants of its offspring *A* and *B* produce the ectoderm of the entire body except the posterior end. The S_2 cell, also called *EMSt* or entodermal-mesodermal-stomodaeal cell, divides into *E* and *MSt*. *E* is the entoderm cell and gives rise to the entire entoderm, which in nematodes is the midgut epithelium. *MSt* divides into *M* and *St*. *M* is the mesoderm cell and produces the body-wall musculature and the cells of the pseudocoel. *St* gives rise to the pharynx. *St, M,* and *E* lie in this order from anterior to posterior in the ventral side of the embryo and are gradually brought into the interior by a process of gastrulation. Here the *E* cell proliferates a chain of cells that becomes the midgut. The offspring of *M* form a strand on either side of the midgut. *St* proliferates a mass of cells that in *Parascaris* passes into the interior by a regular process of stomodaeal invagination but in other nematodes there is no or little trace of such a process and the stomodaeal mass is brought into the interior during gastrulation, later connecting with the surface ectoderm anteriorly, and the midgut posteriorly. After these processes have occurred, a frontal section will show the embryo to consist of an ectodermal epithelium enclosing a central column consisting anteriorly of the stomodaeal mass, and posteriorly of the midgut cylinder with a strand of mesoderm to either side. The germ layers are thus clearly defined at this time. In the meantime divisions have continued in the *P* line of cells (propagative or germ-cell line). P_2 divides into S_3 and P_3 and P_3 into S_4 and P_4. S_3 and S_4 are generally ectodermal, furnishing the epidermis of the rear part of the body and of the proctodaeum, but in *Parascaris,* their offspring contribute to the mesoderm. This contribution, which may be regarded as an ectomesoderm, adds itself to the rear end of the previously formed mesoderm, which is clearly an entomesoderm, and functions like the latter. The cell P_4, which meantime is brought into the interior by the gastrulation process and lies behind and below the midgut, is the source of the reproductive tract except its distal ectodermal parts. In *Turbatrix* in which the fate of P_4 has been followed more precisely than in other nematodes, P_4 divides into two cells, S_5 and P_5. The offspring of S_5 produce the epithelium that covers the gonad and forms the wall of the gonoduct and P_5 proliferates the germ cells. Similar observations were made for *Rhabdias* by Neuhaus and the common origin of both germ cells and epithelial wall of the reproductive tract from P_5 probably obtains throughout nematodes. The nervous system and sense organs are formed from the same cells that furnish the anterior epidermis, probably S_1 as a rule. Little definite information is available as to the origin of the excretory canals but in *Turbatrix,* which has a one-celled renette, this is one of the offspring of S_1, or, in other words, the renette is an epidermal gland cell.

As the fate of the individual cells has now been followed, it remains to consider the more general embryonic processes. Cleavage results in a coeloblastula which, however, is generally dorsoventrally flattened and elongated in the anteroposterior direction. The dorsal and lateral walls of the blastula consist chiefly of the descendants of S_1 with some additions from S_3 and S_4, whereas the flat or even concave ventral surface is composed of *EMSt* and P_4 or their offspring. Gastrulation occurs mostly by epiboly; the ectodermal sides grow down and push the ventral surface into the interior. As thc embryo is of elongated shape, the blastopore takes the form of an elongated groove that soon closes over completely. The invaginated material forms an oval mass extending anteroposteriorly in the blastocoel and consisting of the stomodaeum, the midgut, and the gonad primordia with the mesoderm to either side. It is clear that the pseudocoel of nematodes is a persistent blastocoel. By the process of gastrulation the previously flattened embryo is altered to a cylindroid shape. It continues to elongate and is soon recognizable as a juvenile worm curved inside the egg membranes.

The development of the pharynx is rather puzzling in view of the fact that this structure is generally regarded as a stomodaeum, i.e., as of ectodermal nature, as evidenced

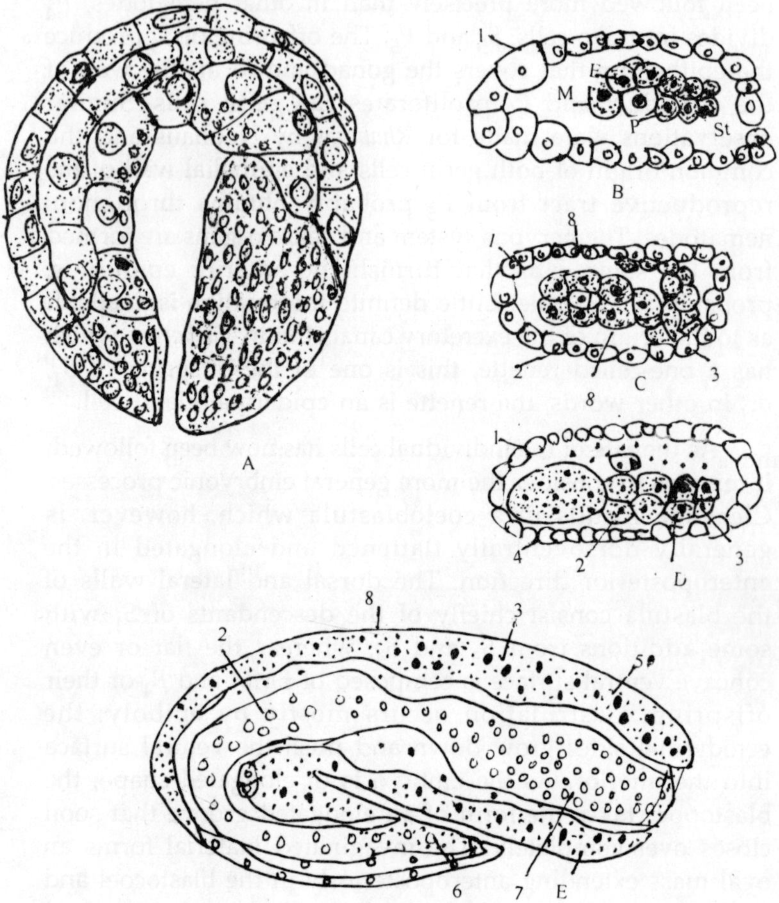

Fig. 1.33 : Embryology (concluded). *A.* Advanced embryo of *Parascaris equorum*. *B–E.* Development of *Turbatrix aceti*, anterior end to right, sagittal sections. *B.* i41-cell stage, after gastrulation. *C.* 171-cell stage, stomodaeum forming, 7 E-cells on each side. *D.* Later stage, stomodaeum formed. *E.* Sagittal section of a young worm, digestive tract completed. 1, epidermis; 2, primordium of reproductive system; 3, stomodaeum; 4, pseudocoelocytes; 5, epidermis derived from S1 (stippled); 6, epidermis derived from S3 (blank); 7, proctodaeum derived from S4 (stippled); 8, midgut, derived from E-cells.

by its cuticular lining. But all the accounts of nematode embryology agree in deriving the pharynx or at least its main primordium from the same stem cell that gives rise to the entoderm and entomesoderm. Some authors do indicate a later participation of ectoderm in the formation of the pharynx as Boveri (1899) for *Parascaris* and Martini (1903) for *Camallanus* but others deny this. However, it must be recalled that the pharynx consists mainly of muscle fibres, presumably mesodermal, and that it epithelial component is demonstrable with difficulty. This fact probably explains the minor role played by the ectoderm in the embryonic development of the pharynx.

It was discovered by Boveri (1887) for *Parascaris equorum* and since confirmed for other nematodes that not only is the germ-cell line segregated from the somatic cells at a very early stage but also this line is distinguishable from other cells by the behaviour of its chromosomes. Only in the germ-cell line do the chromosomes reappear of normal size and shape at each mitosis; in the somatic cells chromatin is continuously cast off and the chromosomes appear in a broken and reduced state. This process of chromatin loss from the nuclei of somatic cells is termed *diminution* and will be found discussed in works on cytology.

Cell multiplication ceases in all the organs except the reproductive system during late embryonic development and the number of cells or nuclei present at that time remains constant throughout the life of the nematode except that amitotic fragmentation of nuclei may occur in the larger species.

Experiment confirms the highly determinate mosaic nature of nematode development. If one blastomere is destroyed in the two-cell stage, the other develops as it would have in a normal embryo and gives little if any evidence of regulatory power.

Postembryonic Development and Life Cycle in General.— Nematodes hatch as juvenile worms, erroneously called larvae by workers on nematodes. These juvenile worms are usually fully developed at hatching except as regards size and the reproductive system together with accessory reproductive structures. At hatching, the gonad usually consists of four cells, two inner ones, products of *P5*, which are the primordial germ cells, and two outer encircling ones, products of *S5*, which give rise to the gonoducts and the epithelial covering of the gonad. During their growth to the adult size, the juveniles undergo some changes of form, especially of the posterior end and of the buccal capsule. A marked alteration of body form is limited to certain females as in the genus *Heterodera*. A remarkable feature of the postembryonic growth of nematodes is the occurrence of molts in which the entire cuticle, including the lining of the buccal capsule, pharynx, rectum, and vagina, is shed. There are probably always four molts and the nematode emerges from the fourth molt as a fully developed adult although growth may continue. In some cases the juvenile worm may enter on a quiescent phase inside the shed cuticle which thus serves as a cyst. In other cases one or two molts may be passed while the young worms are still within the egg membranes so that the juveniles after hatching appear to undergo fewer than the usual number of molts. This is the case with many of the Filarioidea which tend to hatch in a somewhat embryonic undeveloped state. Here the young worms, called microfilariae, are enclosed in a sheath while still within the egg membranes. Although the nature of this sheath is disputed the sheath is probably the cast-off cuticle of the first molt so that on hatching the microfilaria is really a second-stage juvenile.

The life cycle of nematodes presents far more variation than is the case with trematodes and cestodes, and all degrees of entoparasitism occur; there are, however, no true ectoparasites among nematodes. The terminology that has grown up with regard to nematode cycles is often objectionable and does not conform to usage in other groups of invertebrates. As already noted, the so-called larvae are really juveniles and will be so designated. Those that have not yet molted are first-stage juveniles, those after the first molt are second-stage juveniles, and so on. The stage at which the juveniles enter the host is termed the *infective* stage. In many cases, the molted cuticle remains as a protective covering around the juvenile worm enabling it to with stand adverse conditions. Such juveniles are said to be *ensheathed* or *encysted;* the former term is considered preferable. Juveniles inside the egg membranes, usually referred to as embryonated or larvated eggs or ova, will here be termed ovic juveniles. From the shape of their pharynx, juveniles may be designated as rhabdiform, filariiform, etc. Details of the cycles of specific nematodes will be given in the account of the orders, and at this point only a general picture of the numerous variations will be presented.

a. *Direct Cycle.*—In free-living aquatic and terrestrial nematodes, juveniles after hatching grow and undergo minor changes as in other invertebrates, except that molts occur as in echinoderids and arthropods. Growth may be more or less uniform or may show sudden increases after each molt.

b. *Early Stage of Phytoparasitism.*—Juveniles may attach to the roots or other parts of plants, piercing the tissue by the spear or other armature of the buccal capsule and feeding on plant sap and tissues, but do not enter the plant.

c. *Later Stage of Phytoparasitism.*—The juveniles penetrate into the plant, living in the tissues, or in cavities where their activities have caused the death of tissues, or in galls evoked by their presence. In these sites sexual reproduction occurs and the juveniles after hatching escape to infect new plants.

d. *Final Stage of Phytoparasitism.*—Whereas in the foregoing cases the nematode often retains the ability to live free in the soil, feeding on plant remains, nematodes in this category, belonging chiefly to the genus *Heterodera,* are obligatory phytoparasites. Here also the juveniles penetrate into plants where they evoke gall formation. The males retain their vermiform appearance but the females alter to a plump, lemon-like form, and eventually become a skin filled with eggs. These cysts released into the soil by the disintegration of the affected part of the plant are very resistant to adverse conditions. They eventually emit juveniles that migrate through the soil to infect new plants. Species of *Heterodera* may require insemination by males or may reproduce parthenogenetically.

e. *Early Stage of Zooparasitism.*—Here the nematode leads a free existence as males, females, and young but the latter on reaching the infective stage require to be transported as ensheathed juveniles by insects to fresh food supplies, before they can continue their development.

f. *Saprophagous Type of Zooparasitism.*—The nematode leads a free existence in the soil but in the lack of adequate organic food the juveniles enter an invertebrate host where they live ensheathed or free without much development until the host dies from other causes, whereupon the juveniles feed upon the carcass and develop to sexual maturity; one or more generations may be passed in the carcass until it is consumed.

g. *Zooparasitism in Juvenile Stages Only.*—Juvenile nematodes develop to advanced stages within an invertebrate host, chiefly an insect, then leave, and develop to sexual maturity as free-living soil inhabitants, for several generations in some species; but eventually an infective stage, either an ensheathed or ovic juvenile, must be taken into the invertebrate host again. This type of cycle is characteristic of mermithids which do not feed in the adult stage.

h. *Combination of Saprophagous Juveniles with Zooparasitic Adults.*—In this type of cycle, the nematodes grow to sexual maturity in the invertebrate host, mate, and produce offspring; these remain in the host until it dies whereupon they feed on the carcass. Third-stage ensheathed juveniles constitute the infective stage which is ingested by the invertebrate host.

i. *Zooparasitism in Adult Females Only.*—Here the juveniles are free-living and grow to sexual maturity in the soil. After mating the males die and the impregnated females penetrate an invertebrate host where they produce a new generation of juveniles.

j. *Combination of Phytoparasitic Juveniles with Zooparasitic Adults.*—Gravid female nematodes in the invertebrate host produce juveniles that enter the reproductive system of the host and are deposited by the latter in the plant tissue that is parasitized by the host larvae (insects). The juvenile nematodes live in and feed on the plant tissue in company with the host larvae, develop to sexual maturity, and the gravid female nematode then penetrates the host larva, remaining therein as the latter metamorphoses to an adult insect.

k. *Combination of Phytoparasitic Adults with Zooparasitic Juveniles.*—Here the juvenile stages occupy the invertebrate host, which is a plantparasitic insect. The juveniles emerge into the plant tissues where they become adult and breed; the juveniles thus produced gain access to the invertebrate host.

l. *Combination of Phyto-and Zooparasitism with Parthenogenesis in the Zooparasitic Phase.*—The offspring of adult female nematodes inhabiting an invertebrate (insect) host develop into parthenogenetic females. The offspring of these mature in the soil or in a plant and after mating, the males die, while the impregnated females penetrate an insect larva.

m. *Combination of Phyto- and Zooparasitism with Parthenogenesis in the Phytoparasitic Phase.*—The young produced by the gravid female nematodes in the female host fly are deposited by the latter in a plant where they develop to parthenogenetic females; several parthenogenetic generations may ensue. Eventually there is produced a gametogenetic generation and the impregnated females penetrate the fly larvae within the plant.

n. *Continuous Zooparasitism without Free Stage.*—Only one nematode is known, the oxyuroid *Probstmayria vivipara* (Ransom, 1907) inhabiting the intestine of horses, that lives generation after generation in the host without the necessity of passing any stage to the exterior.

o. *Typical Zooparasitic Cycle.*—The nematode gains access to the invertebrate or vertebrate host in a young stage, develops to sexual maturity, usually in the intestine but also in other organs, and its eggs or ovic embryos or juveniles or free young pass to the exterior where they exist for varying periods and may undergo further development. Eventually, however, an infective stage is reached which must enter new individuals of the same host species either by active penetration or through being ingested. After such entry the young worm may proceed directly to the site parasitized by the adult or may undergo more or less extensive wandering through the host body, especially through the lungs, before settling down to become a sexual adult. This cycle may be regarded as an alternation of parasitic and free stages.

p. *Alternation of Zooparasitic Hermaphroditic or Parthenogenetic Females with Free-living Adults.*—The offspring of the parasitic females on reaching the exterior may or may not, apparently depending on conditions, develop into one or more free-living generations of sexual adults. This cycle is considered an alternation of homogonic and heterogonic generations; it occurs in *Strongyloides stercoralis* and other Rhabdiasoidea.

q. *Optional Use of Transport Host.*—Infective juveniles may be ingested by various invertebrates in which they encyst without further development, and reach maturity only when the transport host is ingested by the appropriate definitive host.

r. *Zooparasitism with Obligate Intermediate Host.*—Here some young stage of the nematode is swallowed by the intermediate host or penetrates it, reaches an infective stage in this host, and develops no further until it gains access to the definitive host.

s. *Zooparasitism with Two Obligate Intermediate Hosts.*—In this cycle, known for *Gnathostoma,* the first intermediate host is *Cyclops,* the second a fish, frog, or snake, and the definitive host a carnivorous mammal.

Histological Peculiarities of Nematodes.—An outstanding anatomical feature of nematodes is their cell constancy; as already noted cell division in general ceases at hatching except in the reproductive system and there is usually no further increase in the number of nuclei. Cell or nuclear constancy has been studied in detail in several nematodes, in *Oxyuris equi* by Martini (1907, 1908, 1916), in *Turbatrix aceti* by Pai (1928), and in *Rhabditis longicauda* by Schönberg (1943). Martini found cell constancy in all structures of *Oxyuris* except the reproductive tract, the midgut, and the epidermis. Pai was unable to determine the number of nuclei in the epidermis or reproductive system but listed a total of 434 for other body parts as follows: 251 nerve cells, one excretory cell, 18 cells in the midgut, 64 in the body-wall musculature, 59 in the pharynx, 5 in the esophagus, 20 in the rectum, and 16 mesenchyme cells in the pseudocoel. In the female *Rhabditis,* Schönberg counted 68 muscle cells in the body wall, about 200 nerve cells, 120 epidermal cells, and 172 cells in the digestive tract. The number of muscle cells in the body wall seems to be remarkably similar for a large number of nematodes, being usually 64, 65, or 68. In the horse ascaris, Goldschmidt (1908-1910) reported a total of 162 nerve cells.

As there is usually no cell multiplication after birth in nematodes despite very considerable increase in size, it follows that many cells become very large and elongated. This is notably the case with the muscle cells of the body wall, the cells of the pseudocoel, the cells of the excretory canals, etc. In the larger nematodes, muscle cells may be several millimetres in length, possibly more than 10 mm. in some cases. There is also evidence a tendency toward the formation of syncytia. Nuclear or cell increase during juvenile phases certainly seems to obtain in some nematodes especially in the digestive tract; thus Moorthy (1938) records for *Camallanus sweeti* a steady augmentation in the number of intestinal cells during postembryonic life from 35 in first-stage juveniles to about 200 in the adult. Whether mitotic division is involved is not stated but it seems more probable that the nuclei multiply by amitosis or fragmentation.

Chapter—2

Classification of Aschelminthes

Phylum Aschelminthes is divisible into one class-Nematoda.

CLASS—NEMATODA

1. Body is slender, cylindrical, tapering towards both the ends.
2. Body is covered with cuticle.
3. Intestine is well formed.
4. Body cavity is not lined with epithelium.
5. Cloaca is absent in female.
6. Male and female reproductive organs are well developed.

The Classification of Nematoda is not stable as they have been studied mainly independently. The scheme given below will be found simple and convenient.

Class Nematoda is divided into two sub-classes:

Sub-class I—Aphasmidea

Sub-class II—Phasmidea

Sub-Class I—Aphasmidea

1. Phasmids or caudal sensory organs are absent.
2. Amphids or anterior sensory organs many.
3. Excretory organs either reduced or absent.

4. Coelomocytes and mesenterial tissue are well developed.
5. Candal alae in females are absent, free-living.

The Sub-class is divided into following orders:

Order (1) Chromadoridea

1. Oesophagus is not elongated but differentiated into three regions.
2. Cuticle smooth or ringed. Ornamented and sometimes with bristles.
3. Amphids spiral.
4. Mainly marine and free-living.

   Examples: *Druconematids, Walichoanolaimus.*

Order (2) Enoploidea

1. Mainly marine and free-living.
2. Six sensory papillae on the lip.
3. Ten to twelve sensory bristles arranged in one or two circlets.
4. Cyathiform (goblet-shaped) amphids.
5. A pair of cephalic slits.
6. Size small to moderate.

   Examples: *Pelagonema, Enoplus, Metonchidiamus.*

Order (3) Doyrlaimoridea

1. Found in fresh-water and soil.
2. Cuticle smooth without bristles.
3. Head with an inner circlet of six and an outer circlet of ten sensory papillae.
4. Odontostylet, the piercing apparatus derived from a modified tooth in the oral region.
5. Amphids cyathiform.

6. Posterior part enlarged.

   Examples: *Dorylaimus, Xiphinema.*

Order (4) Mermithoidea

1. Anterior end with 16 labial papillae.
2. Sensory bristle absent.
3. Cuticle is smooth.
4. Amphids either rudimentary or cyathiform.
5. Oesophagus opening in blind intestine.
6. Larvae parasites in invertebrates whereas adults are free-living.

   Examples: *Mermis, Paramermis.*

Order (5) Trichiuroidea

1. Intestinal muscles, caudal gland or suckers are absent.
2. Mouth without lips.
3. Pharynx very long and narrow.
4. Females with one ovary.
5. Males with one or no copulatory spicule.

   Examples: *Trichinella, Trichuris.*

Order (6) Desmocolecoidea

1. Small sized with heavily ringed cuticle.
2. The head is armoured. Anterior end with four sensory bristles.
3. Amphids are crescent-shaped or pump-shaped.
4. They are marine and free-living.

   Examples: *Tricoma, Desmoscolex.*

Order (7) Monohysteroidea

1. Small forms with smooth or ringed cuticle having bristles.

2. Amphid circular.
3. Anterior end with 4, 6, 8, or many sensory bristles.
4. Free living forms, few are marine, some are fresh water and others terrestrial.

   Examples: *Monohystera, Siphonolaimus.*

Sub-class II—Phasmida

1. With phasmids and pore like amphids.
2. Excretory organs with paired lateral canals.
3. Coelomocytes 4-6.
4. Caudal or hypodermal glands absent.
5. Reproductive system well developed.

The sub-class has been divided into following orders:

Order (1) Rhabditoidea

1. Mouth without lips and with 6, 12, or 18 papillae.
2. Cuticle is ringed or smooth.
3. Pharynx, long with posterior bulb.
4. Single ovary in females and males with copulatory spicules.
5. Mostly parasite on birds and mammals.

   Examples: *Rhabditis, Diplogaster.*

Order (2) Oxyuroidea

1. Pinshaped small or medium sized.
2. Cervical papillae absent.
3. Pharynx with posterior bulb.
4. Females with slender tail.
5. Males with one or two copulatory vesicles and sometimes with alae forming a bursa.
6. Parasites in invertebrates and vertebrates.

   Examples: *Enterobius, Oxyuris.*

Order (3) Ascaroidea

1. Large parasitic.
2. Mouth surrounded by three rarely six lips.
3. Cervical papillae present.
4. Pharynx with or without valve.
5. Female short tail.
6. Male with ventrally coiled tail and with two equal copulatory spicules and usually without alae.

   Example: *Ascaris.*

Order (4) Strongyloidea

1. Mouth without distinct lips but has leaf crowns.
2. Pharynx is without bulb.
3. Males are provided with a distinct caudal bursa which is supported with rays.
4. Females with a ovejactor.

   Examples: *Ancylostoma, Necator, Strongylus.*

Order (5) Dracunculoidea

1. There are no definite lips or buccal capsule.
2. Mouth is surrounded by ring of papillae.
3. Pharyngeal bulb is absent. Pharynx is muscular anteriorly but glandular posteriorly.
4. Copulatory bursa is absent in males.
5. They are parasite of vertebrates.

   Examples: *Dracunculus, Micsopleura.*

Order (6) Spiruroidea

1. Mouth with two lateral lips, sometimes 4 or 6.
2. Pharynx with bulb with anterior muscular and posterior glandular part.
3. Males without bursa but have spirally coiled tail.

4. Parasitic forms.

   Examples: *Gnathostoma, Spiroxys.*

Order (7) Rhabdiasoidea

1. Cuticle smooth.
2. Definite pharyngeal bulb is absent.
3. Parasitic stage is hermaphrodite or parthenogenetic free-living stage may develop into males and females.

   Example: *Rhabdias.*

Order (8) Filariodea

1. Slender filiform worms.
2. Mouth simple, lips are absent.
3. Females larger having an anterior vulva and a long tubular vagina.
4. Males with two unequal spicules. Alae present or absent.
5. Found in blood or lymphatic system. Transmitted by blood sucking insects.

   Examples: *Wuchereria, Microfilaria, Loa.*

TRICHURIS

Phylum	—	Aschelminthes
Class	—	Nematoda
Sub-class	—	Aphasmida
Order	—	Trichuroidea
Genus	—	*Trichuris*

1. It is commonly celled as *whip-worm*. It is endoparasite found in the caecum and large intestine of man.
2. It has world wide distribution but common in tropical countries.

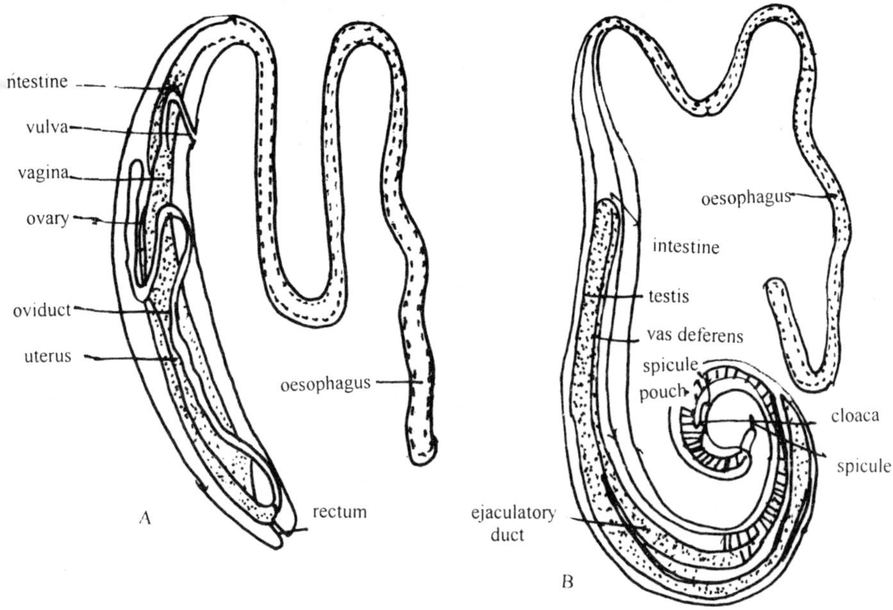

Fig. 2.1 : *Trichuris*

3. Female is slightly larger than male.
4. The anterior part of the body having pharynx is usually more slender and this portion is buried in the mucosa of host's alimentary canal.
5. Mouth is simple without lips.
6. It feeds on blood and epithelial cells.
7. Life-cycle is simple with no intermediate host.
8. Development of eggs take place in moist soil. Embryos develop in about three weeks.
9. Contaminated food and water responsible for infection in man by swallowing eggs with juveniles.
10. Liberated juvenile grow into adults within a month.
11. Excretory system is absent.
12. Heavy infections is responsible for diarrhoea and anaemia.

TRICHINELLA

The systematic position is the same as that of *Trichuris.*

1. It is parasite of man, pig, rat and other vertebrates. It is commonly known as *trichinia worm*.
2. It is commonly found in Europe, United States and Arctic regions.
3. Sexes are separate, males are about 1.5 mm. in length and the females are 3 mm. in length.
4. In both the sexes, the mouth is without lips.
5. Pharynx is long, slender and non-muscular and is embedded in stichosome.
6. The oesophagus is tubular, well developed and occupies 1/3 of the total length of body.

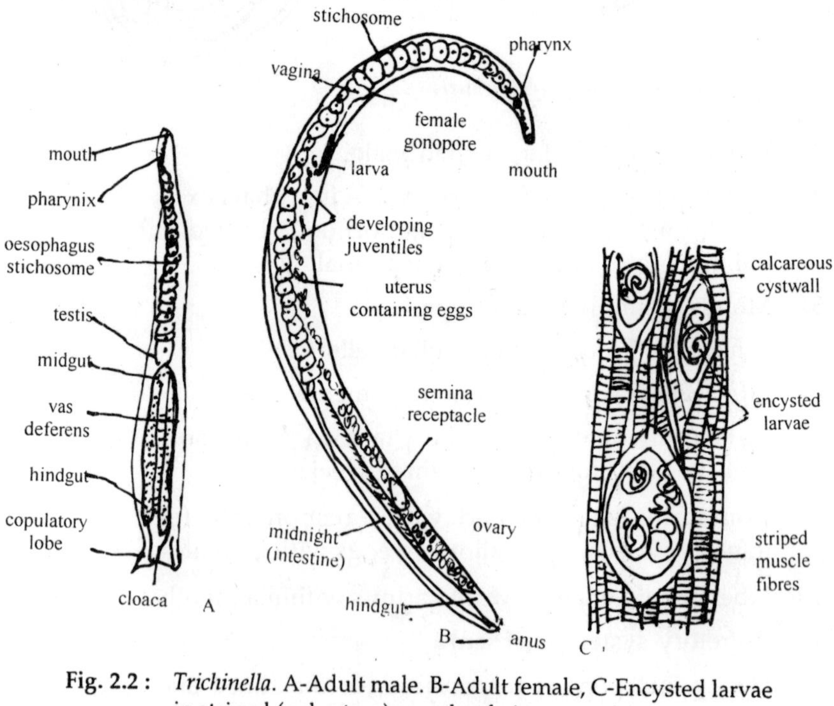

Fig. 2.2 : *Trichinella*. A-Adult male. B-Adult female, C-Encysted larvae in striped (voluntary) muscle of pig.

7. The female worms which are viviparous enter the lymphatic vessel and deposit their young ones. Each young worm then reaches the muscular system where they secrete a cyst around them. This is called encysted sexless stage.
8. The encyated sexless trichinae are ingested by man along with pork-flesh, where the cyst wall is dissolved by the action of digestive juice and the reproductive organs start developing.
9. The developing worm begins to migrate into the muscles and produces the symptoms of a dreaded disease called *Trichinosis*.
10. Patients feel gastro-intestinal pain, and weakness with or without fever.

RHABDITIS

Phylum	—	Aschelminthes
Class	—	Nematoda
Sub-class	—	Phasmida
Order	—	Rhabditoidea
Genus	—	*Rhabditis*

1. Genus *Rhabiditis* includes numerous free-living, semi-parasitic forms. They are commonly found in organic matter, faces of man. They feed on the fluid formed by bacterial decying of organic matters.
2. The pharynx has two bulbs with an isthmus joining them.
3. The excretory system is H-shaped.
4. The reproductive system is J-shaped.
5. The large female has a pointed tail and didelphic sex-organs i.e. having paired ovaries, oviducts, seminal recepticles and uteri.

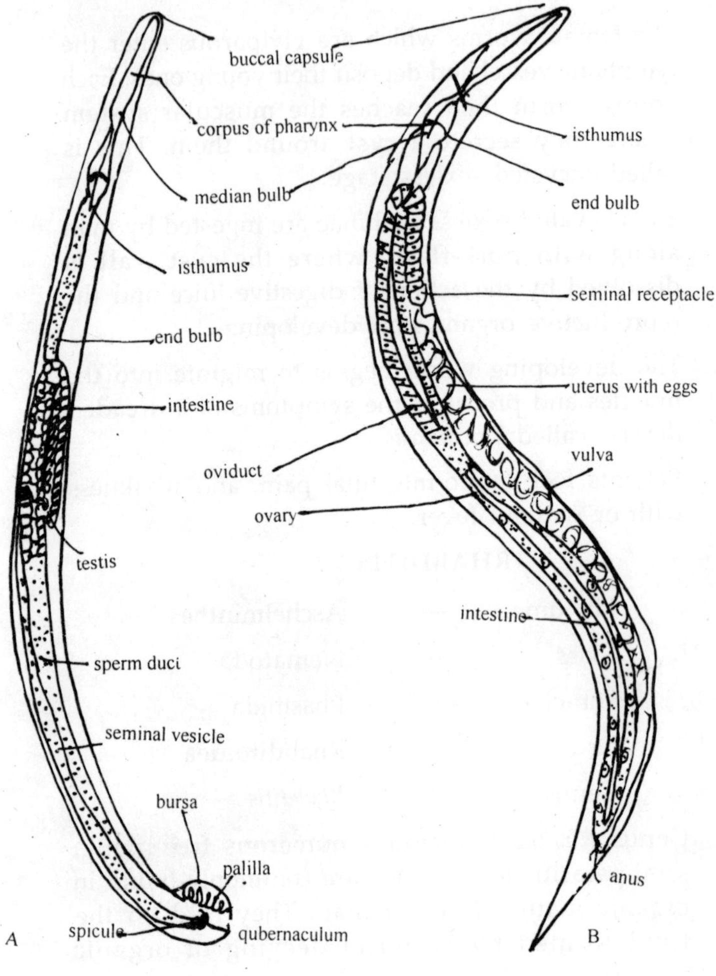

Fig. 2.3 : *Rhabditis* A = male, B = female.

6. The male has a flattened copulatory bursa, two equal spicules and monorchic sex organs i.e. single testis.
7. Life cycle is simple and direct. The juveniles adhere to the bodies of dung beetles for transport to fresh dung.

ENTEROBIUS VERMICULARIS

Phylum	—	Aschelminthes
Class	—	Nematoda
Sub-class	—	Phasmida
Order	—	Oxyuroidea
Genus	—	*Enterobius*
Species	—	*vermicularis*

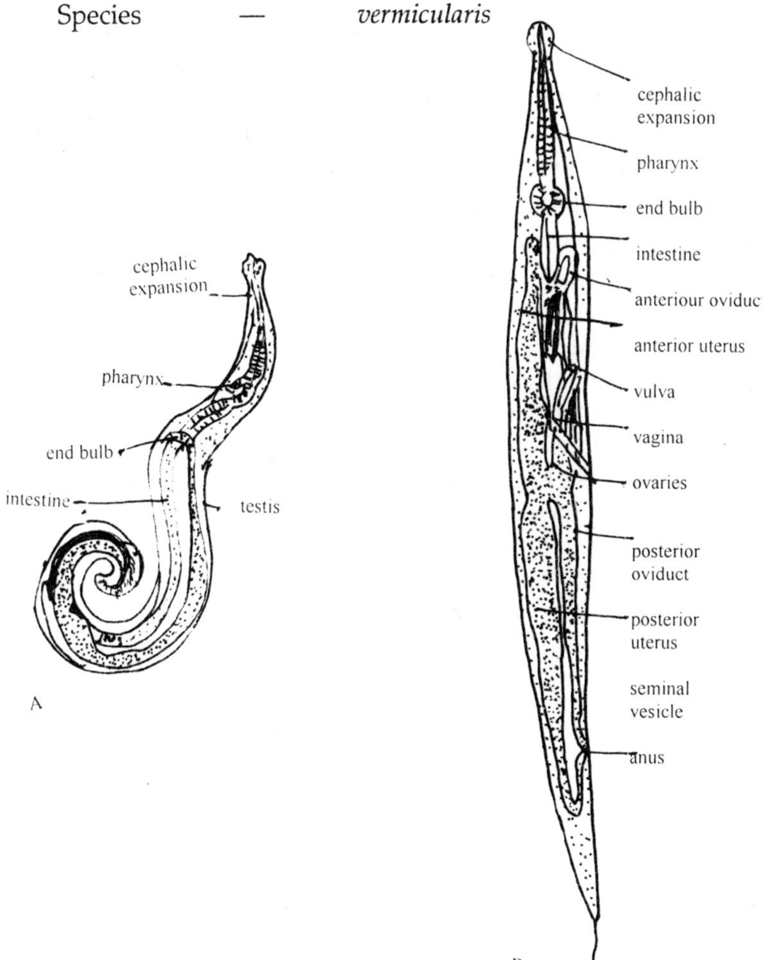

Fig. 2.4 : Entrobius A = male, B = female.

1. It is commonly called *'pinworm'* or *'seatworm'*.
2. It has world-wide distribution, but specially found in Europe, India, Africa, Canada, Chile and U.S.A.
3. Adult worms live in the calcum, appendix and adjacent parts of intestine of man.
4. Sexes are separate. The male measures 2-5 mm. in length and the female measures 8-13 mm. in length.
5. Anterior end is provided with three lips and a pair of cephalic expansion.
6. Male has a ventrally curved tail with a single spicule and a pair of cuticular expansions known as caudal alae. It is monorchic.
7. Female has a straight tail which becomes long and pointed. It is didelphic Vulva is anterior.
8. Life-cycle is simple there is no intermediate host Gravid females migrate to anus causing itching sensation. They crawl and deposit eggs on the perineal skin. Due to scratching, gravid females rupture and thus small eggs are liberated. The reinfection in children takes by putting fingers in mouth. Eggs hatch. Juveniles inhabit small intestine. They moult twice and turns into adults which reach their normal habitat. The infection is called *Enterobiasis*. Symptoms are inflammation of caecal mucosa, abdominal pain, insomnia, hysteria, restlessness, loss of appetite and sometimes appendicitis.

LOA LOA

Phylum	—	Aschelminthes
Class	—	Nematoda
Sub-class	—	Phasmida
Order	—	Filarioidea
Genus	—	*Loa*
Species	—	*loa*

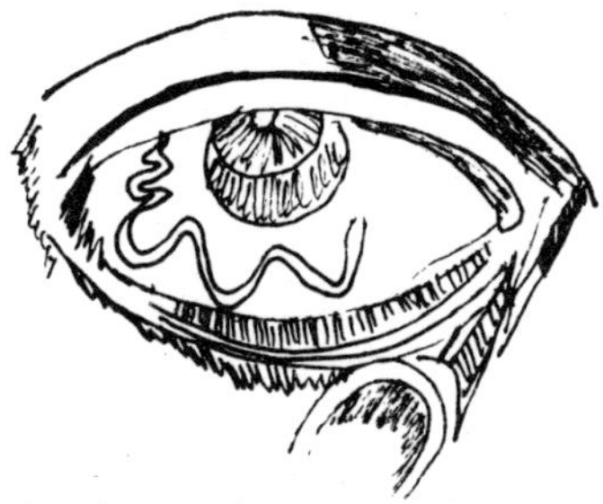

Fig. 2.5 : Loa loa in the cornea of man.

1. It is commonly known as *African eyeworm,* chiefly found in Africa.
2. They live in the subdermal connective tissue of man in West Africa.
3. The male and female measure 20-35 mm. and 20-70 mm. in length respectively.
4. Body is covered by numerous warts.
5. The intermediate hosts are mango flies, *Chrysops.* The larvae moult and develop up to infective stage in fly. These are transmitted in blood of man through proboscis.
6. The adult worm moves about in the body in the subdermal tissue and sometimes passes across the eye ball causing intense itching and swelling known as 'calabar swellings'.
7. These are injurious and fatal when they penetrate brain and spinal cord.

DRACUNCULUS

Phylum	—	Aschelminthes
Class	—	Nematoda
Sub-class	—	Phasmida
Order	—	Dracunculoidea
Genus	—	*Dracunculus*
Species	—	*medinesis*

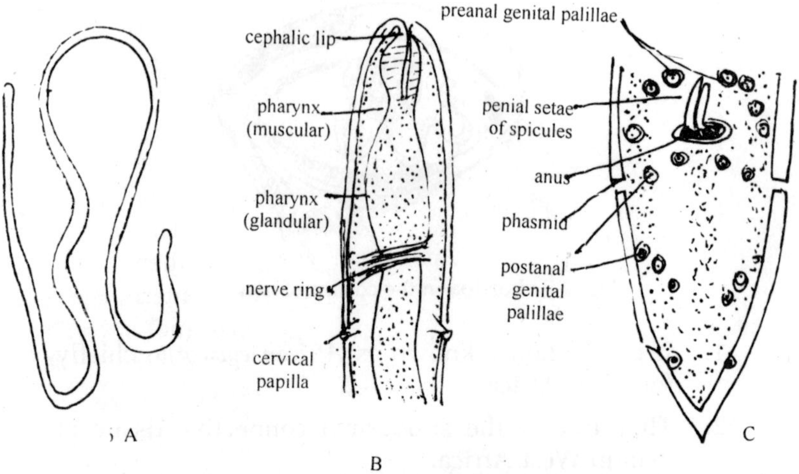

Fig. 2.6 : *Dracunculus medinesis.* A-Entire. B-Anterior end of male. C-Posterior end of male in ventral view.

1. It is commonly called *'Guinea Worm'*. It is found in tropical countries like India, Africa, South-America and West Indies.
2. It lives in the deeper subcutaneous tissues of man.
3. The anterior end is blunt white the posterior end is pointed.
4. The adult male measures 4 cm. in length. It's hind end is curved and bears 10 pairs of genital papillae and a pair of pineal spicules.
5. Adult female measures 70-120 cm. in length. It's hind end is straight and bears a spine-like process.
6. The oesophagus is distinguished into an anterior narrow muscular region and a posterior broader region glandular region.
7. The female reproductive system is didelphic with opposed uteri.
8. Females are viviparous. The alimentary canal and vulva are atrophied in mature females.

9. Mature female migrates to the superficial layers of the skin in those parts of the body which are likely to come in contact with cold water. Toxic substances are produced under the skin which results in the formation of a blister, leading to an ulcer, which breaks on coming in contact with water. On coming in contact with water, the worm ruptures, releasing larvae in water. The larvae swims about but do not develop further unless swallowed by *Cyclops,* which serves as intermediate host. Here the larvae moults twice and become infective. Infection occurs when *cyclops* are swallowed by a new host alongwith drinking water.
10. It causes *guinea worm disease* characterised by asthma, nausea, vomiting, diarrhoea etc.

Chapter—3

Ascaris Lumbricoides

Much is known about the morphology and life-cycle of Nematoda but out of them *Ascaris* is most known. *A. lumbricoides* is the most common nematode parasite found in the human intestine. Being of large size and found abundently, *Ascaris* is studied as the main example of the nematodes.

SYSTEMATIC POSITION

Phylum	—	Aschelminthes
Class	—	Nematoda
Sub-class	—	Phasmida
Order	—	Ascaroidea
Genus	—	*Ascaris*
Species	—	*lumbricoides*

Habit and Habitat

A. lumbriocoides, a large nematode parasite of man, is world-wide in distribution. It occurs more commonly in the small intestine of children than in adults because of the children's custom of playing in contaminated, egg-laden soil adjacent to dwellings. As many as 1000-1500 adult worms may be present in a single host. Because of their strong, pointed and cuticulate bodies and their habit of wandering, they may perforate the gut wall and cause peritonitis or they may enter the bile or pancreatic ducts or the vermiform appendix and interfere with normal digestive functions.

External Features

Shape and Size. Both the sexes are elongated, cylindrical and pointed at both the ends. Males 15-31 cm. long and 2-4 mm. in diameters, while the females are 20-35 cm. long and 4-6 mm. in width. The males are further distinguished by a ventrally curved posterior end, from the slit-like and opening of which often protrude two copulatory spicules.

Colouration. In the living condition, the colour of the animal is milky-white or reddish-yellow. There is a characteristic lustre on the body.

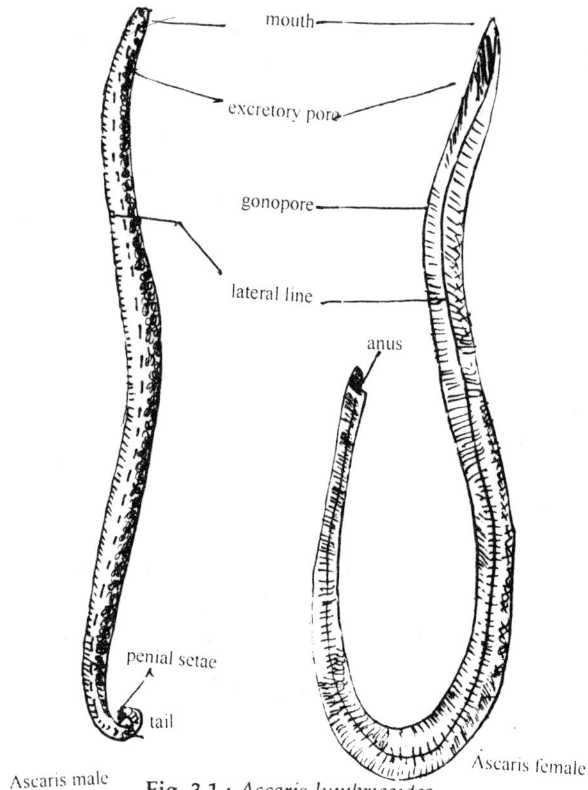

Fig. 3.1 : *Ascaris lumbricoides.*

Longitudinal lines. There are four longitudinal lines or streaks, running throughout the length. One is mid-dorsal, one mid-ventral and two lateral. The dorsal and ventral lines are white in colour, the later streaks are thicker and broader than dorsal and ventral streaks and are brown in colour.

Mouth. A triangular mouth is present at the anterior end of the body. It is surrounded by three denticulated lips. One of these is mid-dorsal and other two are latero-ventral in position. The mid-dorsal lip is almost elliptical and has a forked fleshy core in the centre. A circle of four *sensory papillae* is found on it. The remaining two lips also have a forked core and two rows of sensory papillae. Each sensory structure is called *amplids.* The sensory papillae of nematodes are very useful in the identification of the species and assignating their position in the classification. The sensory papillae are *tangoreceptors.* The amphids are *chaemoreceptors.*

The posterior end of the male bears ventrally, 50 pairs of *pre-anal papillae* in front of the cloacal aperture and five pairs of *post-anal papillae* behind it. The first two pairs of post-anal papillae are double. These papillae of the male help in copulation.

Excretory pore. It is situated mid-ventrally, a little behind the mouth.

Genital pore. It is present only in female and is situated mid-ventrally at about 1/3 distance from the anterior end.

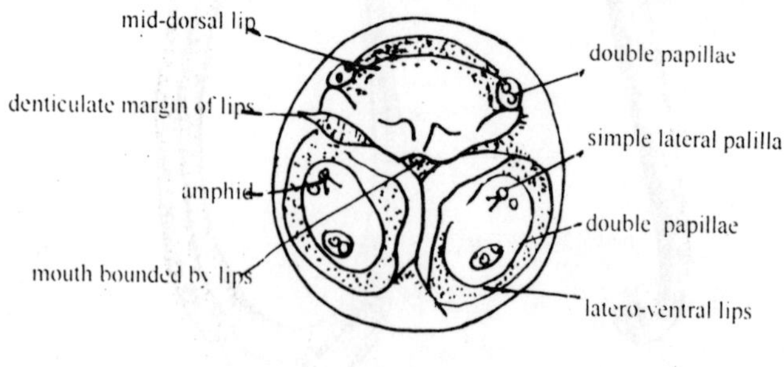

Fig. 3.2 : *Ascaris.* En-face view of lips and mouth.

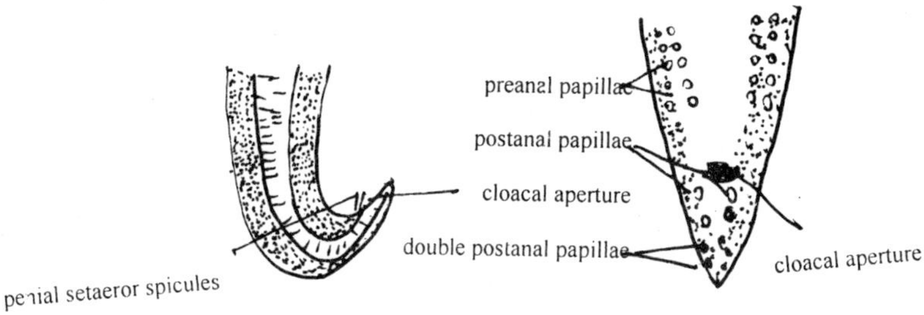

Fig. 3.3 : *Ascaris.* Posterior end in lateral vie and ventral view.

Anus. It is also present only in female. It is a transverse pore quarded by thick lips, lies mid-ventrally a little infront of posterior end.

Cloacal aperture. It is present only in male. It is situated mid-ventrally in front of the curved posterior end. Two chitinous *penial setae* or *penial spicules* are protruding through cloacal aperture.

Body Wall

The body wall is made up of three layers:

(1) Cuticle

(2) Epidermis

(3) Muscle layer

1. *Cuticle.* The cuticle forms a thin transparent elastic but resistant covering around the body. It protects the body against the action of digestive juices is secreted by underlying epidermis. Histologically, the cuticle is formed of several layers and can be distinguished into (i) Cortical layer, (ii) Matrix, (iii) Fibrous layer and (iv) Basement membrane.

   (i) *Cortical layer.* It is made up of dense keratin. It protects the animal from the digestive juices of host.

   (ii) *Matrix.* It is an elastic layer containing sulphur rich matricin.

(iii) *Fibrous layer.* It is made up of collagen fibres that cross each other obliquely.

(iv) *Basement membrane.* It forms the inner limit of cuticle.

Under electron-microscope the cuticle is made up of following layers:

(1) *Lipoid layer.* It is in the form of a thin osmophilic membrane, about 1000A° thick.

(2) *Cortical or cortex layer.* It consists of (a) *outer cortical layer* which lies as discontinous strips or rings around the body, and (b) an *inner cortical layer.*

(3) *Matrix layer.* It consists of (a) an outer *fibrillar layer* traversed by branching canals, (b) a *homogeneous layer* which several radial striations, and (c) a *boundary layer* resembling the outer cortical layer.

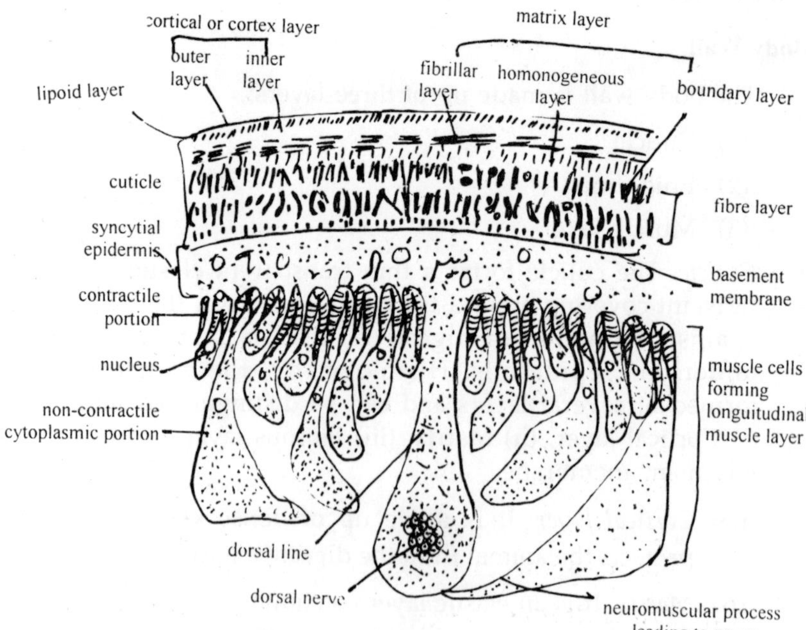

Fig. 3.4 : Body wall in section.

(4) *Fibre layer*. It consists of collagen fibres arranged in three strata.

(5) *Basement membrane.* It is a thin layer surrounding the epidermis.

1. *Epidermis.* It is *syncytial* in nature as it is having no cellular boundaries i.e. the nuclei are scattered in cytoplasm. It is thickened and projects into the pseudocoel as four longitudinal ridges called, *epidermal chords*. These are dorsal, ventral and lateral in position. The lateral chords are thicker than others, containing longitudinal excretory canals and lateral nerves in them. The dorsal and ventral are thin containing nerve cords and terminate towards the posterior end of the body.
2. *Muscle layer.* Beneath the ectoderm is a single layer of longitudinal muscles which is distinguished into four quadrants. Of them two are dorso-lateral and two ventro-lateral. Each containing about 150 muscle cells. The muscle cells are highly specialised spindle-shaped

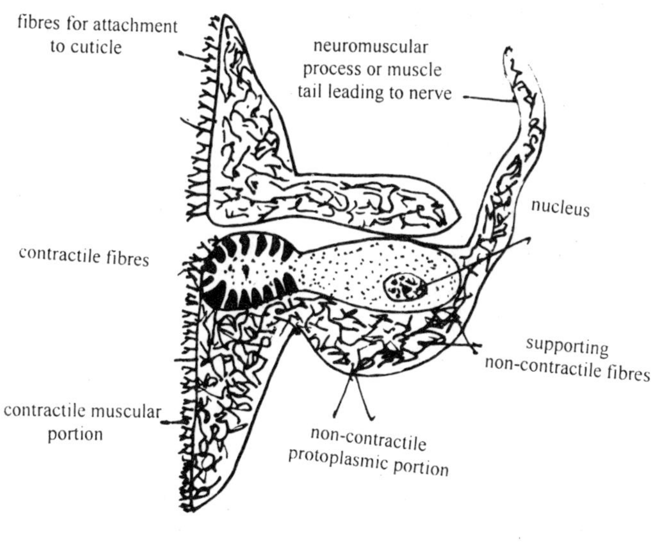

Fig. 3.5 : *Ascaris*. A muscle cell in T. S.

cells, distinguished into two parts lying below the ectoderm:

(a) *Fibrillar region.* Formed of contractile and non-contractile fibres arranged alternatively, it is found wrapped around the protoplasmic part.

(b) *Protoplasmic region.* It is the inner part of the muscle cell situated towards the pseudocoel. It is almost bladder-like in appearance and consists of a single nucleus.

There is a complete absence of circular muscles in nematodes.

Pseudocoel

There is a large fluid filled space between the body wall and the viscera. The fluid contains fibrous tissue in the

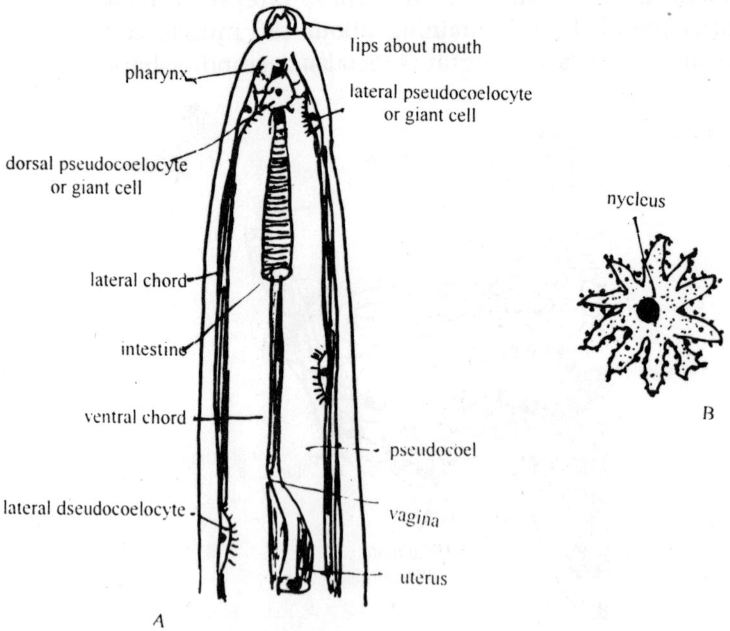

Fig. 3.6 : *Ascaris.* A-portion of giant cells in pseudocoel. B-single giant cell.

form of fenestrated membranes and fixed cells. The epithelial lining of this cavity is absent hence cannot be a true coelom. The fenestrated membranes extend over the viscera and muscle layer from a single giant *pseudocoelocyte* placed on the dorsal side where pharynx connects with the intensine. Four more star shaped giant cells lie in relation to epidermal cords. Thus in all there are five pseudocoelocytes. The pseudocoelocytes and fenestrated membranes may be regarded as a kind of mesenchyme.

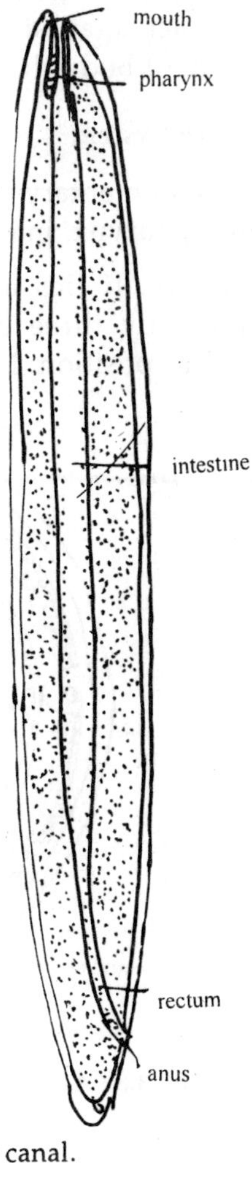

Fig. 3.7 : *Ascaris.* Alimentary canal.

In fact, the spaces are said to be large vacuoles in a few, enormous cells. A protein-rich fluid of characteristic, unpleasant odor fills the cavity. This fluid may produce some of the toxic and inflammatory symptoms of ascarid infections.

Locomotion

Locomotion is by means of longitudinal muscles which produce undulating movements of the body in the dorsoventral waves. The circular muscles are absent to counteract against longitudinal muscles, the hydrostatic pressure of body fluid against the cuticle serves the purpose.

Digestive System

The alimentary canal of *Ascaris* is complete. It comprises mouth, pharynx, intestine, rectum and anus.

Mouth. It lies at the anterior end of the body. It is bounded by three lips—one mid dorsal and two latero-ventral in position. It leads into the pharynx.

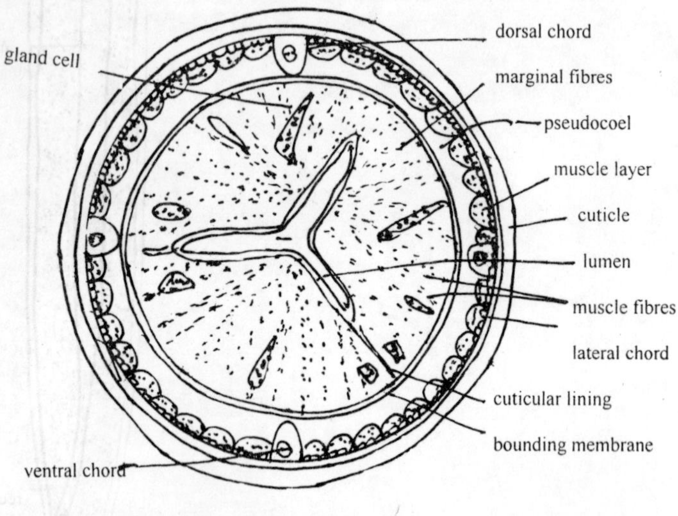

Fig. 3.8 : *Ascaris.* T. S. body passing through pharynx.

Pharynx. The pharynx is muscular, about 10 mm. long. It provided with a triradiate, cuticulate lumen, and has a wall composed of radially arranged, lumen-dilating muscles. A single, pinnately branched dorsal, and a pair of palmately branched subventral, uninucleate glands are embedded in the wall of the pharyns. These are unicellular structures discharging their contents into phanyngeal cavity. The wall is composed of syncytial epithelium. The pharynx is followed by the intestine. The pharynycal wall projects into the intestine as a phanyngeo-intenstinal valve.

Intestine. It is a long, thin, walled, dorso-ventrally flattened tube which is devoid of muscles. The intestinal wall is formed of a single layer of elongated columnar cells lined externally by a *basement membrane* and a thin layer of *cuticle.* The free inner margins of columnar cells are produced into several finger-like projections called the *microvilli.* These form a *brush border* and increase the surface area for absorption.

Rectum. The intestine is followed by a short *rectum* which is also dorso-ventrally flattened. Wall of the rectum consists of tall columnar cells and is lined internally by cuticle and externally by muscle tissue. In male, the rectum opens into the *cloaca* which also receives the ejaculatory duct, while in female, it opens out through the anus.

Anus. The anus is a transverse slit guarded by anterior and posterior lips and is provided with a special muscle, the *depressor ani.*

Digestion. Ascaris gets semi-digested or digested food from its host. So it does not require to digest food. However, it can digest semi-digest food by means of enzymes secreted by gland cells of pharynx and cells of anterior part of intestine. These enzymes can digest carbohydrates, proteins and fats.

*Absorption.*The cells of intestine generally absorb digested food. Excess fat and glycogen are stored in different tissue.

Egestion. By the contraction of dilator muscles attached to rectum undigested foot is expelled out.

Respiration

There being no special organs of respiration, but the parasite carry on anaerobic respiration by breaking down glucose and forming carbon dioxide and fatty acid which are eliminated out through the cuticle. In the presence of oxygen, it respires aerobically. The small amount of haemoglobin in the body wall (muscles) and pseudocoel fluid takes up the oxygen even when it is present in low tension.

Excretory System.

The excretory system of *Ascaris* is built entirely on a different plan. The flame cells or nephridia are absent. Renette cells of osmoregulatory function found in other nematodes are not found here.

In *Ascaris,* the excretory system is made up of a syncytium containing four or more nuclei. It consists of:

1. *Excretory duct.* There is an excretory duct having two or more nuclei embedded into its walls. The excretory duct opens out through excretory pore.

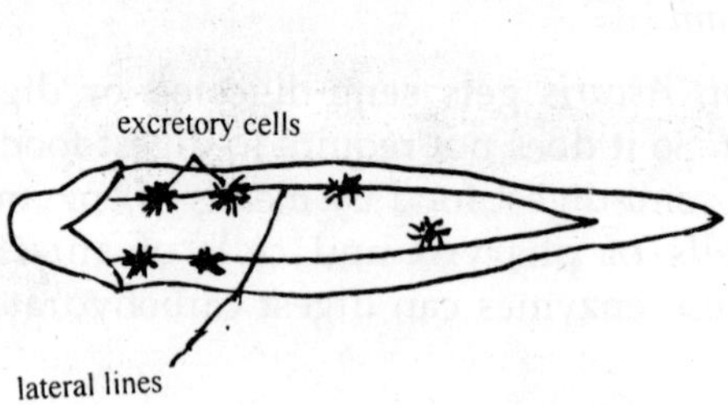

Fig. 3.9 : *Ascaris.* Excretory system.

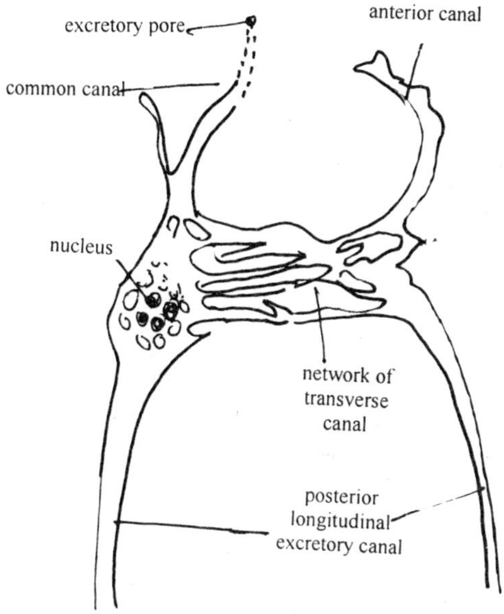

Fig. 3.10 : *Ascaris.* H-shaped excretory system in anterior region of body.

2. *Excretory cell.* There is an excretory cell comprising two lateral excretory filaments running along each side of the body adjacent to the lateral chord.
3. *Commissure.* An excretory bridge is formed where two lateral filaments join. Each excretory filament contains an intracellular excretory canal. The intracellular excretory canals divide up into a network of canaliculi in the commissural region. Excretory products are believed to diffuse into the canals which secrete urea, which is eliminated out through the excretory pore.

Nervous System

The nervous system consists of ganglia and nerves, all in close association with the ectoderm. The nervous system of *Ascaris* has been thoroughly studied by *Goldschmidt*. The number of cells taking part in the formation of entire system is apparently fixed and it is 162. There is a distinct *circumpharyngeal nerve ring* around the pharynx which connects a pair of *lateral ganglia,* a *dorsal ganglion,* a pair of *subdorsal ganglia* (each sub-divided into six smaller ganglia) and a pair of large *ventral ganglia* from these ganglia of the nerve-ring there proceed backwards eight nerves: a *mid-dorsal nerve* that passes through the dorsal epidermal chord; a *mid-ventral nerve* that passes through the ventral epidermal chord; a pair of *dorso-lateral nerves* that run just above the lateral chords; a pair of *lateral nerves* the extend through the lateral epidermis chords and a pair of *ventro-lateral nerves* that run just below the lateral epidermal chords. Of these, the ventral nerve is most prominent and is ganglionated through out its length, therefore, it is regarded as nerve

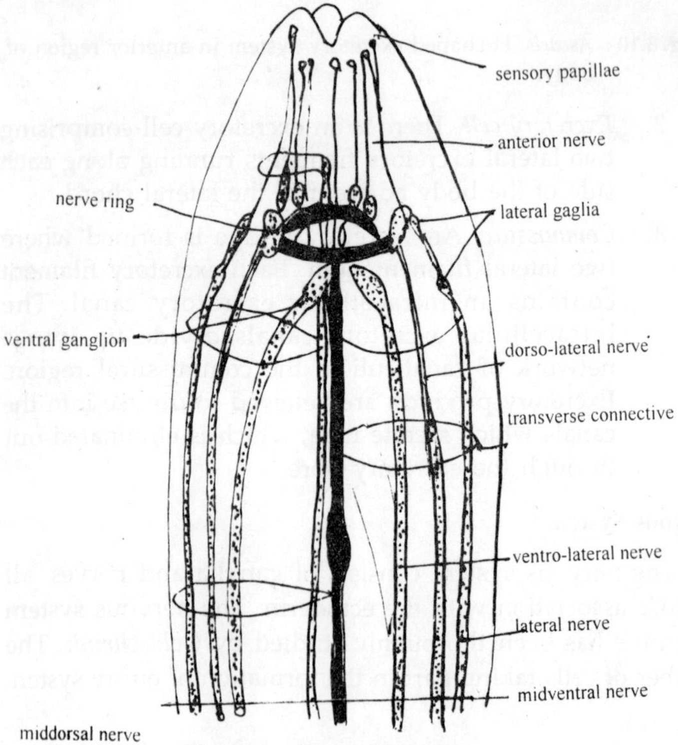

Fig. 3.11 : *Ascaris.* Nervous system.

cord. Posteriorly the nerve cord joins a *pre-anal ganglion*. The dorsal and ventral nerves are connected at intervals by asymmetrical *dorso-ventral connectives* throughout their length. Similarly, the ventral and lateral nerves are also connected at interval by *ventro-lateral commissures*.

One of the six ganglia formed by sub-division of each lateral ganglion is known as *amphidial ganglion* which supply nerves to amphids of its side. The genital papillae of the male worm are innervated from special *bursal nerves*.

Sense Organs

The sense organs include papillae amphids and phasmids.

Papillae. Various types of papillae such as *labial, cervical* and *genital papillae* are present in *Ascaris*. A labial papilla consists of a nerve axis placed inside a cuticular elevation surrounded by supporting cells. The dorsal lip has two double papillae, each consisting of a labial papilla on the medial side and a cephalic papilla on outerside. Each ventro-lateral lip has a labial papilla near its upper margin and a double papilla near its lower margin. The double papilla a labial papilla towards the medial side and a cephalic papilla

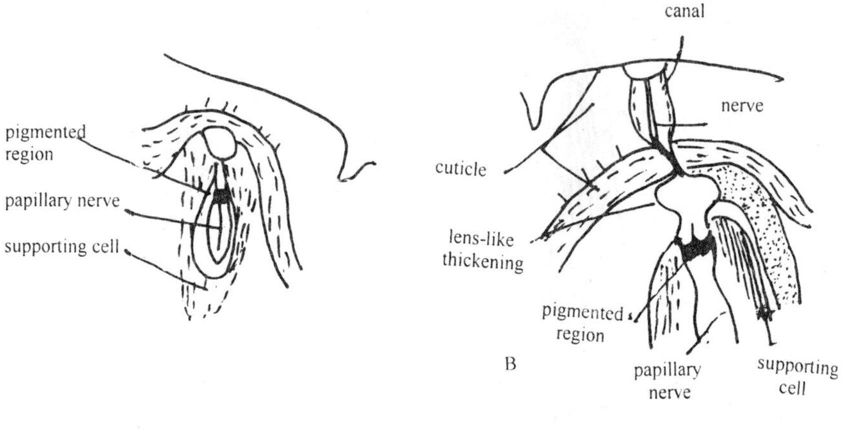

Fig. 3.12 : *Ascaris*. Caplatic and labial papillae.

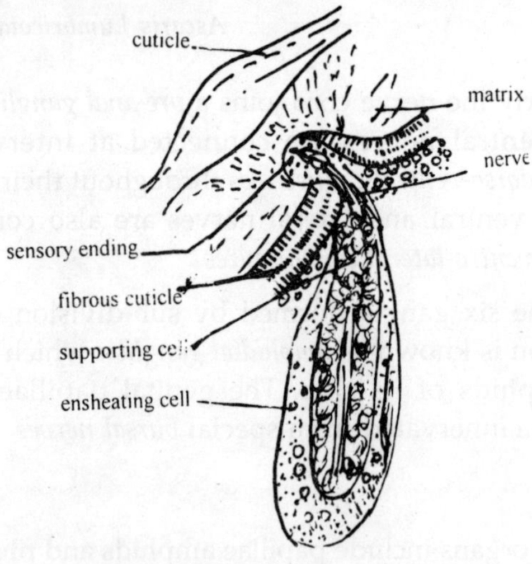

Fig. 3.13 : *Ascaris*. Cervical papilla.

towards the outer side. Thus the lips bear six labial and four cephalic papillae in all. Genital and caudal papillae have a similar structure.

Amphids are two in number, one on each ventro-lateral lip a little above the labial papilla. An amphid consists of

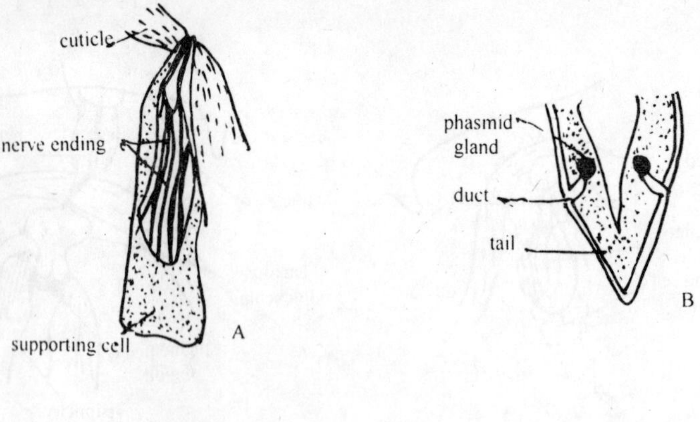

Fig. 3.14 : *Ascaris*. A-amphid. B-phasmid.

supporting cell opening outside by a pore and having a bundle of nerve-fibres that unit into one fibre. These are probably a chaemoreceptor.

A pair of *phasmids* occur on each side of the tail. Phasmids is the out let of a unicellular gland opening over a small papilla which is probably chaemoreceptor.

Reproductive System

Sexes are separate and there is a well distinct sexual dimorphism.

I. Male Reproductive System

The male reproductive organs are:

1. Testis
2. Vas deferens

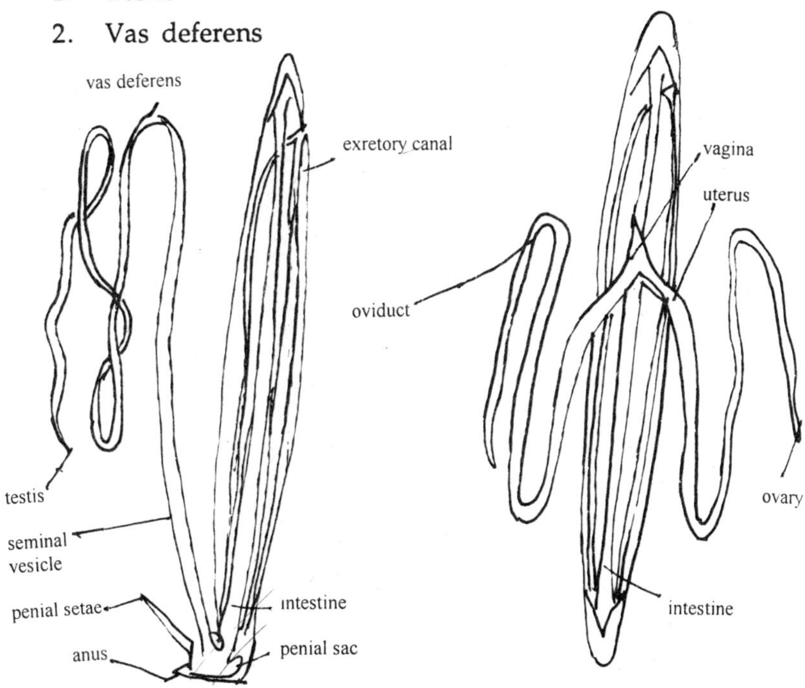

Fig. 3.15 : *Ascaris*. A-male reproductive system. B-female reproductive system.

3. Seminal vesicle
4. Ejaculatory duct
5. Penial sac with penial setae.

1. *Testis*. The testis is a long thin, coiled and thread-like structure in the anterior part of pseudocoel. It represents *monorchic condition* (presence of single testis) and *telogonic* mode of origin of germ cells (*i.e.* germ cells developing only from the proximal part of gonad). The testis can be distinguished into three parts:

   (a) anterior solid part consisting of sex cells, (b) middle part and (c) posterior part. The central axis of the tube is in the form of a solid *cytoplasmic rachis*, around which the clusters of amoeboid sperms in various stages of development.

   It is continued posteriorly into the vas deferens.

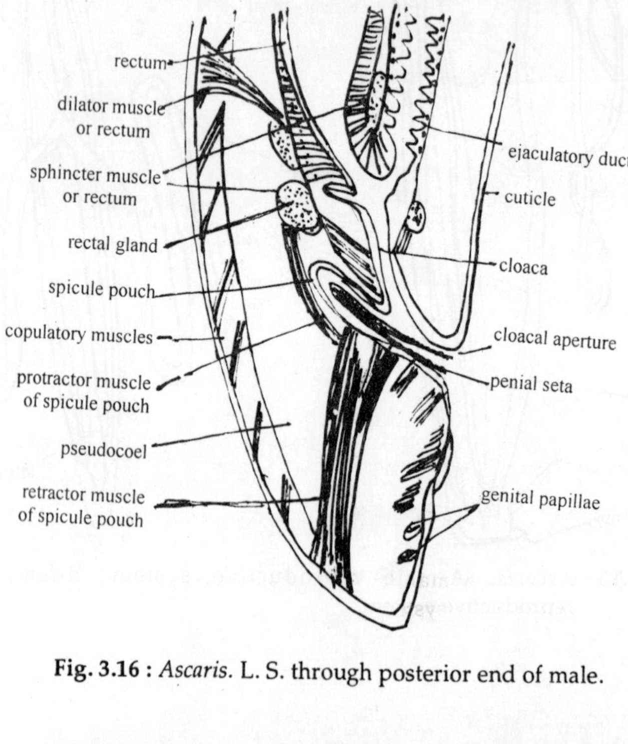

Fig. 3.16 : *Ascaris*. L. S. through posterior end of male.

2. *Vas deferens.* The vas deferens is as wide as the testis and the two cannot be marked externally. It is very short and leads into the seminal vesicle.
3. *Vesicula seminalis.* The vesicula seminalis is a wide tube present in the posterior 1/3rd part of the body. It joins the ejaculatory duct.
4. *Ejaculatory duct.* It is a short, narrow but muscular tube opening into the cloaca, which leads outside by cloacal opening.
5. *Penial sac with penial setae.* On the dorsal side of the cloaca are a pair of sac-like structures called penial sac. These are formed by evaginations of the cloaca. Each pouch secretes and lodges a club-shaped *spicule* or *penial seta*.

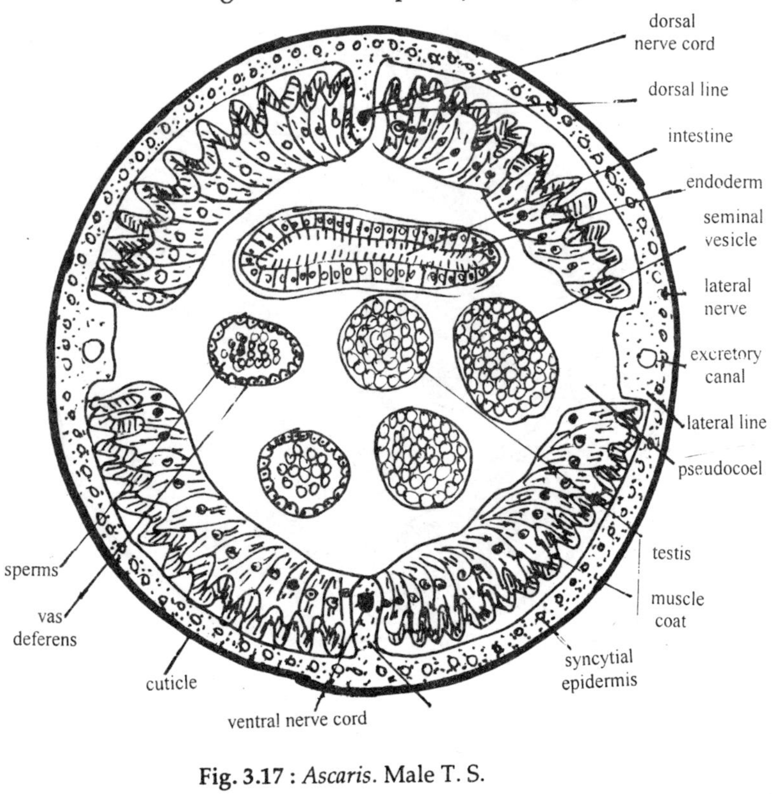

Fig. 3.17 : *Ascaris.* Male T. S.

The spicule is enclosed in a sheath and consists of a cytoplasmic core surrounded by thick cuticle. Each spicule measures 2-3.5 mm. in length. The penial sac have special *protractor* and *retractor* muscles from their walls, the muscles push the spicule out and withdraw them back into the pocket.

II. Female Reproductive Organs

The female reproductive organs include:

1. A pair of ovaries
2. Oviducts
3. Uteri
4. Vagina

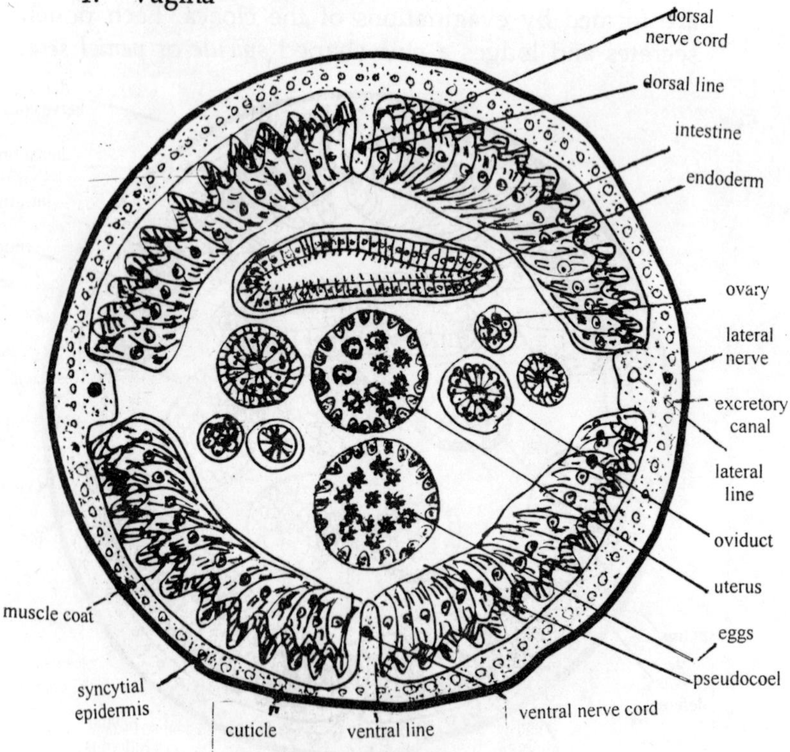

Fig. 3.18 : *Ascaris.* Female T. S.

1. *Ovaries*. The ovaries represent *didelphic* condition i.e. there are two ovaries. Each ovary is a long, thread-like, much twisted and blind tubule occupying posterior 1/3rd of the body. Its walls consists of a single layer of cuboidal epithelial cells lined externally by basement membrane. There is no lumen, the central part is occupied by the cytoplasmic *rachis*, around which young eggs are clustered.
2. *Oviducts*. The oviducts are two thin ducts very similar to the ovaries. These open into the uteri.
3. *Uteri*. Each oviduct continues into a much wider, thicker and almost untwisted tube, the *uterus*. The wall of the uterus is composed of tufted cells, surrounded by a muscle layer with inner *circular* and outer *oblique* fibres.
4. *Vagina*. Two uteri join to form one vagina which opens to the outside by female genital pore or *vulva*, lying on mid-ventral side about 1/3rd of the length of the body posterior to anterior end.

Life-cycle

Ascaris is *monogenetic* i.e., it requires only one host to complete its life-cycle.

1. *Copulation and Fertilization*. *Ascaris* copulate in the intestine of man. During the process sperms are ejected by male into vagina of the female. These move in the uterus where fertilization occurs. While moving further the eggs become covered by a chitinous shell and an outer irregular *albuminous coat*.

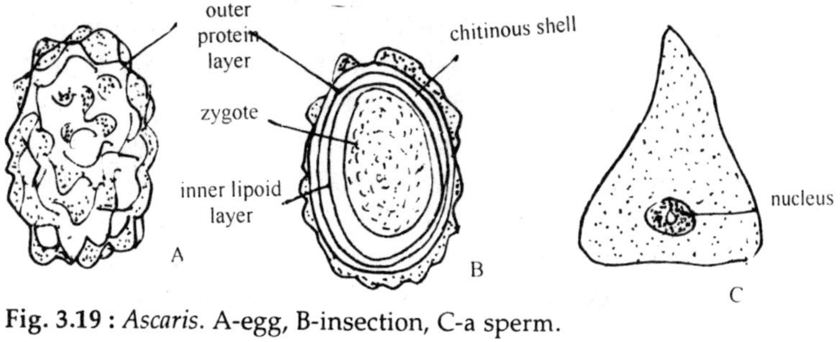

Fig. 3.19 : *Ascaris*. A-egg, B-insection, C-a sperm.

The eggs are round or oval (60μ × 45μ), light brown in colour and float in saturated solution of common salt. The unfertilized eggs laid by the mother may mix up with the fertilized eggs but the unfertilized eggs are larger (90μ × 45μ) and sink in saturated salt solution.

2. *Egg-laying*. The uteri of a single female may contain as many as 2,70,000,00 eggs (*Cram*, 1927), with an average daily production of about 2,00,000 eggs. The fertilized eggs are released in the intestine of host through the vulva. From the intestine of host they are eliminated along with faecal matters.
3. *Development*. The development takes place at 85° F i.e. outside the body of man. It is stopped at 60° F and egg gradually degenerated at 100° F. They can remain alive for years in moist soil.

Cleavage. The cleavage is spiral and determinate. The first division is transverse producing an upper *dorsal cell* (*AB*) and a lower *ventral cell* (*P*). The dorsal cell (*AB*) divides vertically into an *anterior cell* (*A*) and a *posterior cell* (*B*). Simultaneously, the ventral cell divides horizontally into *upper cell* (*EMST*) and *lower cell* (*P*). Thus the four-celled embryo has *T*-shaped appearance. The P_2 cell shifts towards right and comes to lie in front of the upper cell (*EMST*). The *A* and *B* cells divide together into right and left cells and by further divisions form the ectoderm of the embryo. The *EMST* and P_2 cells divide into *E, EMST*, P_3, and *C* cells. The offsprings of cell E form the endoderm, while the descendants of *EMST* constitute the mesoderm and a part of ectoderm. The cell P_3 divides into P_4 and *D*. The cells *C* and *D* divide and contribute to the ectoderm and mesoderm. The cell P_1 divides into G_1 and G_2 and forms the primordial germ cells. Thus blastula stage is attained. It undergoes gastrulation by invagination and finally develops into an active juvenile, representing the *rhabditoid stage*. Under favourable conditions of temperature, moisture and oxygen supply, it requires 10-14 days to reach the 1st stage larva. After a week or so it moults inside the egg shell and changes into the second stage larva. The hatching and further development of the larva is possible only if the egg enters the final host.

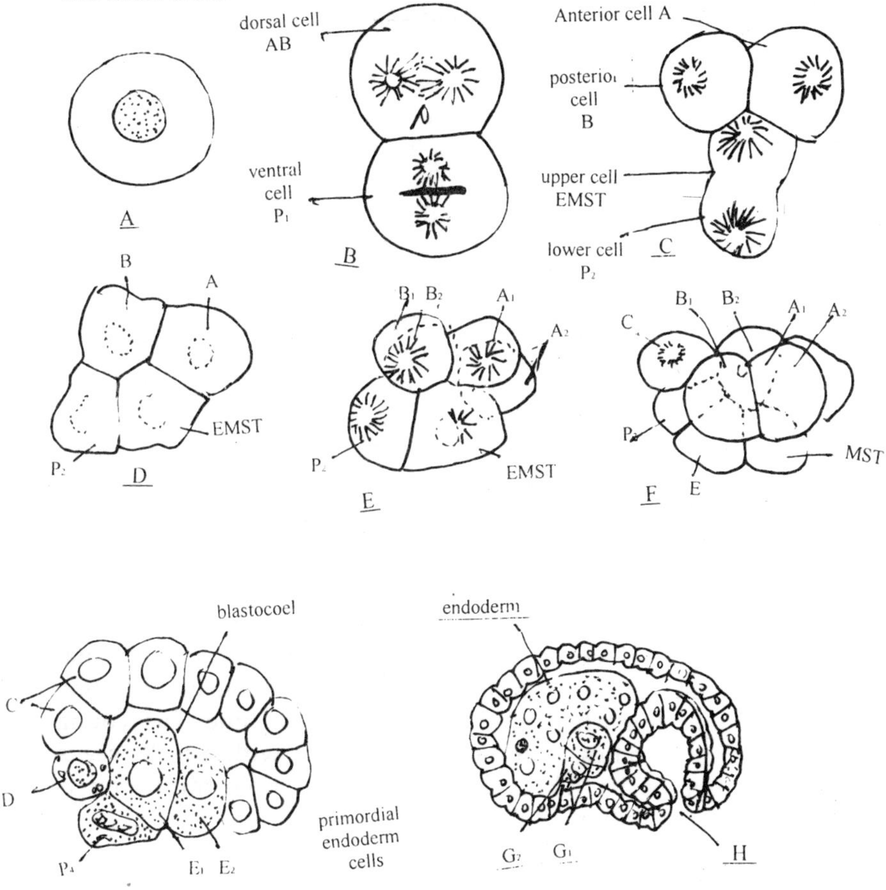

Fig. 3.20 : *Ascaris*. Stages showing early development. A-Zygote, B-2-cell stage, C-4-cell stage (T-shaped), D-4-cell stage (rhomboidal), E-6-cell stage, F-8-cell stage, G-Median sagittal section through blastula, H-Median sagittal section through the embryo after invagination of stomodaeum and the primordial germ cells.

Infection and Emergence of Larva. The eggs containing rhabditiform larva contaminate food and water. If someone takes this contaminated food or drinks contaminated water the eggs reach the healthy man. This type of *transference* is called contaminative transference or infection. The shell of the egg is weakened in oesophagus and is dissolved in the intestine of the new host. The larva is about 200-300µ in length and 14µ in breadth. It has fully formed alimentary canal, excretory and nervous system.

Primary Migration of Larva. The larva burrows the mucous membrane of the intestine of the host and through hepatic portal vein reaches the liver. From here through portal vein and post-caval reaches the right auricle of the heart. From here through pulmonary arteries reaches lungs. In lungs it moults twice. The first moulting takes place after 5-6 days (third stage larva) and second moulting after 10 days forming fourth stage larva. This larva now increases in size and is about 2 mm. in length.

Secondary Migration. From lung the larva reaches bronchi, trachea, larynx and finally to pharynx where it irritates causing coughing by which it re-enters the alimentary canal when the cough is swallowed. After reaching the intestine through stomach it moults fourth or last time. The young Ascaris becomes adult sexually mature within 6-10 weeks and begins its life-cycle again. The whole migration is completed within 10 days.

Pathogenesis

Ascaris is pathogenetic to man. When larvae migrate through the lungs, they cause lung-inflammation and fatal pneumonia. Sometimes, *fever, anaemia, leucocylosis* (presence of excess of white corpuscles in the blood) and eosinophilia result. In mature form, *Ascaris* may or may not disturb life-activities. Some abdominal discomfort and acute colic pains with vomiting, *diarrhoea* and *slight temperature* occur. Often masses of Ascaris block the intestine, in which condition they can only be removed by surgical operations. Several hundred thousands of worms are responsible for blockade and they are fatal. Appendicitis is also caused by the blocking of appendix by Ascaris. Toxic substances produced by *Ascaris* cause *convulsions,* (an involuntary contraction of the involuntary muscles of the body), *delirium* (insanity or disorder in mind), *coma* and *general nervousness. Sang* (1922) found that toxins produced by *Ascaris* cause interference with the digestion of proteins.

Therapy

Several drugs like *Santonin Antepar, Chenopodium* and *Hexylresorcinol* are used effectively under physician's directions. A mixture of oil of *chenopdium tetrachlorothylene* is usually successfully used. *Lamson* discovered that *hyxylresorcinol* is also very effective particularly when given as *crystoids* (in *gelatin* capsules) with fasting for 12 hours before treatment and 4 hours afterwards followed by sodium sulphate for removal of deal worms.

Hetrazan syrup is also very effective and nontoxic especially for children. *Piperazine hydrate,* Alcopar and Banocida are also recommended by several workers as antihelminthes.

Prevention

Adequate sanitation plus education should prevent ascariasis. Lack of adequate fertilizer in many parts of the undeveloped world means that night soil must be used to fertilize crops. Due to the resistance of *Ascaris* eggs to standard disinfectants, the situation would appear to be hopeless. There is a realistic solution through; *Thitasut* (1961) reported that aqueous solution containing 120 ppm free iodine killed *Ascaris lumbricoides* eggs within 5 min or less and did not effect the flavour of leafy vegetables.

One of the problems in control in preventing children from ingesting eggs with contaminated soil given the habits of children throughout the world, including dirt eating, playing in contaminated soil and then placing their fingers in their mouths, ascariasis will unfortunately be with us until adequate sanitation is accomplished.

Chapter—4

Ancylostoma

No group of the nematodes causes more injury to man or greater economic loss through attacks on his domestic animals then the members of the order Strongyloidea, the great majority of which are parasites of mamamals. Many are blood suckers and cause severe injury to their hosts by loss of blood sucked by them or wasted from hemorshages, the result is anemia.

SYSTEMATIC POSITION

Phylum	—	Aschelminthes
Class	—	Nematoda
Order	—	Strongyloidea
Genus	—	*Ancylostoma*

Distribution

Human hookworm is endemic in warm, wet climates, but each species has some limitation on its distribution.

Ancylostoma duodenale is the common hookwork of man in tropical, subtropical and temperate regions of Asia, Africa; Pacific islands and South America. The other important human hookwork is *Necator americanus,* more common in the United States. Some other species of *Ancylostoma,* namely, *A. caninum* and *A. braziliense,* both in dogs and cats, and *A. malayanum* in bear, are also common.

Habit

A. duodenale is the principal hookworm that infect humans. Following early morning, they are found in the small intestive lumen.

In this chapter we will consider only a few hookworms, but the principles apply to many other species in the group.

Name of Organism and Disease Caused

Ancylostoma caninum Hall, 1913, causes hookworm disease, ancylostomiasis, and hookworm anemia. Additional species of hookworms are found in dogs but they will be considered here only incidentally.

Hosts and Location

These hookworms are found in a variety of canids and other carnivores; they are the more prevalent hookworm of dogs. Adult worms occur in the small intestine of the definitive host; the larvae migrate from skin through the blood-stream to the lungs before reaching the gut.

Distribution and Importance

They are cosmopolitan in warm and temperate climates throughout the world; prevalence is high in warm climates, but infections are found in cold to temperate areas such as the northern United States and southern Canada. If we take some of the data for dogs in North America and arrange then from warm to cooler climates, we can see that the prevalence generally decreases toward the north: Hookworms sometimes mature in abnormal hosts, and hosts other than man can be infected with human hookworms.

Morphology

The name 'hookworm', is rather misleading because there is absence of hooks. The anterior hook-like bend probably is responsible for the common name, hookworm.

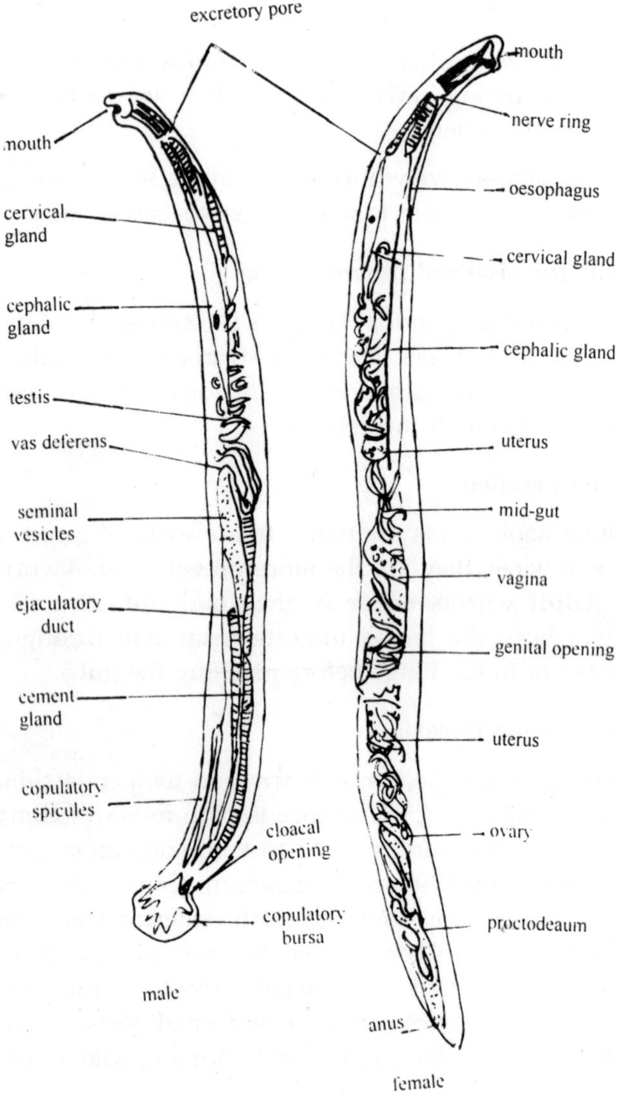

Fig. 4.1 : *Ancytostoma duodenale.*

Shape and Size. The sexes are separate. It is a cylindrical worm and whitish or grey in colour. The females are longer than males measuring 10-13 mm. in length and 0.6 mm. in width. The males measures 8-11 mm. in length and 0.45 mm. in width.

Buccal capsule. The anterior end of male and female is similar, somewhat narrower and curved dorsally. It bears a large cup-shaped buccal capsule for attachment to the intestinal all. The buccal capsule has two *chitinous plates,* bearing teeth. The *teeth* make a tiny cut in the intestinal wall of the host and hook-worm sucks blood by the sucking action of pharynx.

Copulatory bursa. The posterior end of female is somewhat taper, bluntly rounded in the form of a cone and terminates into a minute spine.

The *posterior end* of male is broad and with an umbrella-like structure called *bursa.* The bursa is supported by fleshy bursal rays of definite arrangement.

Digestive System. The digestive system is a straight tube comprising mouth, buccal capsule, pharynx, intestine, rectum and anus.

The pharynx has triradiate lumen lined with in turned cuticule and a wall composed of strong radiating muscle fibres extends posteriorly from the buccal capsule about 1/6 the body length. The posterior region of pharynx is slightly dilated and embedded in its wall are three pharyngeal glands. One is median dorsal in position and a pair located subventrally. These glands open into the pharyns near the nerve ring. An intestive begins immediately behind the triradiate pharyngeal valve and proceeds as a straight tube to a short rectum. It is composed of a single layer of columnar cells. In the intestive the blood of host and other tissues are digested.

Musculature. The somatic muscles consists of a single layer of cells underlying a hypodermis and arranged into four longitudinal groups. These four groups are interrupted by four longitudinal quadratal chords located on the dorsoventral and lateral axes. Because muscle cells are limited to two in each sectioned quadrant, the worms are called *meromyarial*. The almost complete lock of transverse muscle cells.

Excretory system. The exretory system is rather complicated. Flame cells are absent in nematodes. The excretory system, consists of two canals embedded in the lateral chords and an isthums in the pharyngeal region. Surrounding or adjacent to the insthums, are other excretory structures including an excretory sinus into which tributaries from the lateral canals open and from which ventrally is given off a short, cuticulate, terminal duct. The duct opens at a ventral excretory pore. From the excretory sinus two large, subventral glands of uncertain function arise posteriorly. The tributory canals emerge from subventral glands and penetrate the lateral chords where they join the lateral canals. These lateral canals extend from the region of buccal capsule to the posterior end of the body whose they end blindly.

Nervous system. Encircling the pharynx, in front of the excretory pore, is a nerve ring. From a large ganglion located on the ventral portion of the nerve rig, a large nerve runs

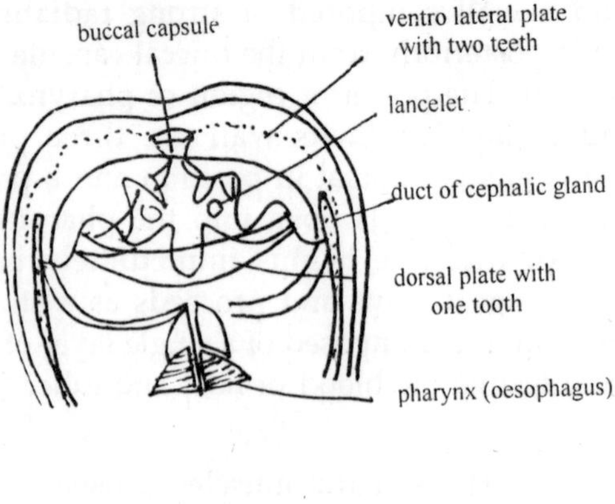

Fig. 4.2 : *Ancytostoma*. Buccal capsule.

in the ventral chord to the posterior end, and smaller nerves run in the lateral chords anteriorly and posteriorly. Four small nerves run forward from the ring to supply anterior end. These main nerve trunks are connected in a complicated fashion by nerve commissures.

Reproductive System

The male and female reproductive system are as follow:

Male reproductive system. The tubular testis occupies the middle third of the body. It begins at the junction of the middle and posterior body thirds and runs in a random fashion to the middle of the body, from which point forward it coils loosely about the intestive. It leads posteriorly into vas deferens followed by seminal vesicle. Seminal vesicle which connects posteriorly to an ejaculatory duct leading directly to the cloaca. The ejaculatory duct, through most of its course, lies between two large, multicellular cement glands whose secretion apparently serves as an adhesive material holding the pair of worms together at copulation.

The posterior end of male bears a wide bursa, two copulatory spicules and a gobernaculum. The bursa is a long, wide, bilaterally. Symmetrical structure, with supporting, paired, fleshy rays having a specific distribution. The two spicules are long and slender. Each measures 2000μ. They are protruded and retracted by special muscles and help in the transport of sperms to the vulva of female. The gubernacculum is a hard piece of on the dorsal wall of cloaca. It is said to be used to guide the spicules.

Female reproductive system. Two much convoluted orarian tubules, one anterior and other posterior to the level of gonopore are present. From the ends of the ovaries nearest the vulva short, narrow oviducts carry the ova into wider seminal receptacles where fertilization occur. From each seminal receptacle arises a muscular uterus that terminate into vagina. Two vagina join and then open to the surface at the vulva.

Life-cycle

Copulation and fertilization. Fertilization is internal and during copulation the bursa of male is applied on the vulva of female. The female Ancylostoma lays about 20,000 eggs per day. The egg is about 60 × 30μ. These are passed along the host faeces at four-celled stage and do not develop further until exposed to air. Under favourable conditions the embryo hatches out of the egg in less 24 hours in the faces contained in the mud.

Larval development. Under favourable conditions of moisture, oxygen supply and temperature (68-85°F) the embryo develops and hatched as larva.

The hatched larva is the first larval stage or the fist *'rhabditiform larva'*. It has mouth, pharynx and a simple intestine. It feeds on bacteria and organic debris in the faeces and grows rapidly. At the end of 2nd day it undergoes first moult and changes into second stage larva (or the *strongyliform larva*). At about 5th day it undergoes second moult to become *filariform-juvenile.* This is the infective non-feeding stage. The skin of the second stage larva is retained as a protective sheat. The larvae live in the upper half inch of soil.

Infection. The third stage or the filariform juvenile represents the infective stage. It may remain alive in the moist soil for nearly four months without feeding. When a man walks bare-footed on the infested soil, the larva

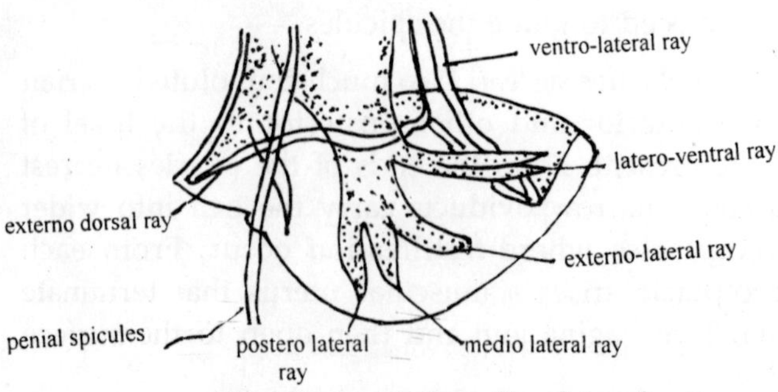

Fig. 4.3 : *Ancytostoma*. Copulatory bursa.

penetrates the skin. Infection can also occur by ingestion of infective juvenile in food or drink.

Larval migration. Within 24 hours of infection, the larval reach the blood-vessels. From there they are carried first to the right ventricle of the nearest and then to the lungs. From lungs the larval assend the trachea and to the pahrynx. In pharynx they are swallowed through oesophagus and finally reach to the small intestive. In the ileum they become attached to the mucous lining and feed on blood of the host. In 5-6 weeks, they moult twice to become adult and begin to reproduce. The normal life-span of hookworm is 5-7 years.

Locality	*Percentage Infection*
Mexico	50-98
Hawaii	71
Georgia	86
Oklahoma	72
Kansas	21
New Jersey	22-27
Illinois	7
Quebec	9

It is difficult to assess the importance of the infection of companion animals either in lost days at work or in production losses, as might be done with some other hosts such as humans or domesticated animals. However, animals do become anemic and undoubtedly do not feel well. There is also some hazard of infections in humans.

Diagnosis

The usual method of determining whether an animal is infected is finding typical eggs in the feces. In areas where hookworm is a continuing problem, it should be suspected when animals become listless and have a rough hair coat and pale mucus membranes indicating anemia.

Some of these clinical signs can be related to other parasitic infections such as *Dirofilaria immitis* or inadequate diet. Therefore, a fecal examination is important to ensure a proper diagnosis.

Host-Parasite Interactions

Damage to the host involves the following:

1. The site of larval penetration

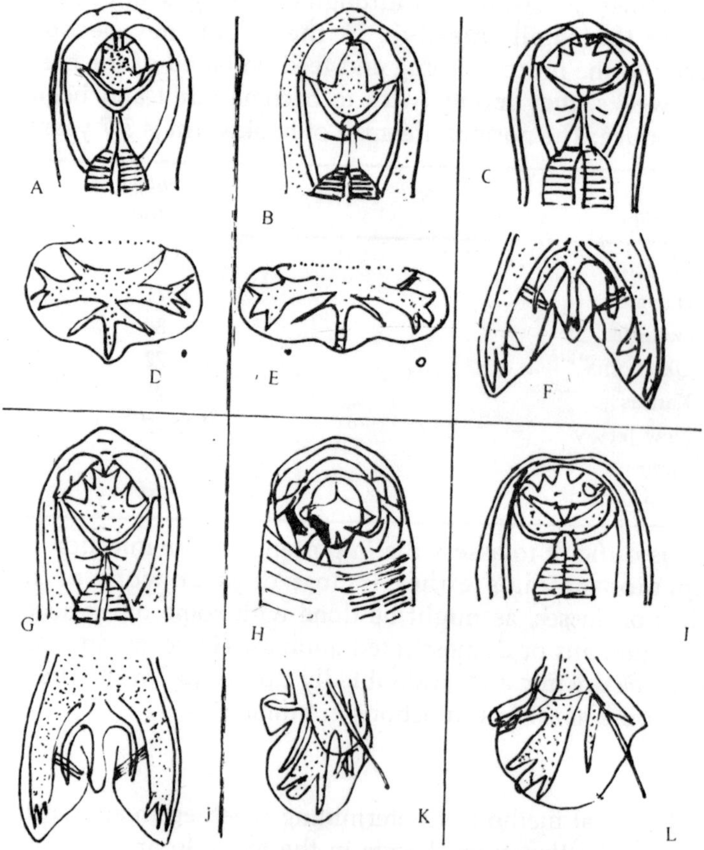

Fig. 4.4 : Hookworm heads and tails. *Ancylostoma ceylanicum* head (*A*) and bursa (*B*); *Ancylostoma braziliense* head (C) and bursa (*D*) *Ancylostoma caninum* head (*E*) and bursa (*F*); *Ancylostoma tubaeforme* head (G) and bursa (*H*); *Necator americanus* head (*I*) and bursa (*J*); *Ancylostoma duodenale* head (*K*) and bursa (*L*).

2. Larval migration
3. The intestine
4. Extraintestinal sites.

The L_3s are about 730 by 22 μ and enter the unbroken skin of the host through the hair follicles. The exact means by which penetration of the skin occurs is still a matter for investigation, but the larvae move down the hair follicle and exit near the basement membrane into the dermis. In the normal host, they soon reach the bloodstream and are carried to the lungs; however, in an abnormal host, hookworm larvae may remain in the subcutaneous tissue and migrate there for some weeks. There is some slight damage to the skin where each larva enters; the greater the number of larvae, the greater the damage.

During migration, the principal site of damage is the lungs, where larvae break out of the capillaries into the alveoli. There may be pneumonia and coughing at this time.

In the intestine the worms moult from L_4 to L_5 and begin to attack the intestinal mucosa. It has been the consensus that hookworms suck blood, but Kalkofen (1970, 1974) concluded that the worms feed on intestinal tissue and that ingestion of blood and blood loss are incidental. When worms feed, they pull a plug of intestinal mucosa into the buccal capsule, the tissue is digested before it passes to the intestine. Digestion required about 10 min and the worm then takes another bite of tissue. Blood is lost when the capillary bed in the subepithelial layer is ruptured; blood flows out around the worm and some is ingested. A number of observers have pointed to the fact that blood ingested by hookworms seems to pass through the gut unchanged; red cells are unaltered and the blood is still bright red when leaving the anus of the worms.

Kalkofen's (1970) view of feeding is clearly at odds with the views held by many investigators and persuasively set forth by Roche and Layrisse (1966) in their exhaustive study of hookworm anemia. They relied particularly on direct observation of feeding hookworms in the open intestine of animals and on *in vitro* studies. They note that (1) large quantities of blood pass rapidly through the gut of worms, (2) worms spew out large volumes of blood when the body wall is nipped with an instrument, (3) there is continual pumping of the esophagus of worms under observation, and (4) the intestine fails to bleed extensively where worms have detached. Finally, it appears that a factor in the blood is required for glycogenesis in the worms.

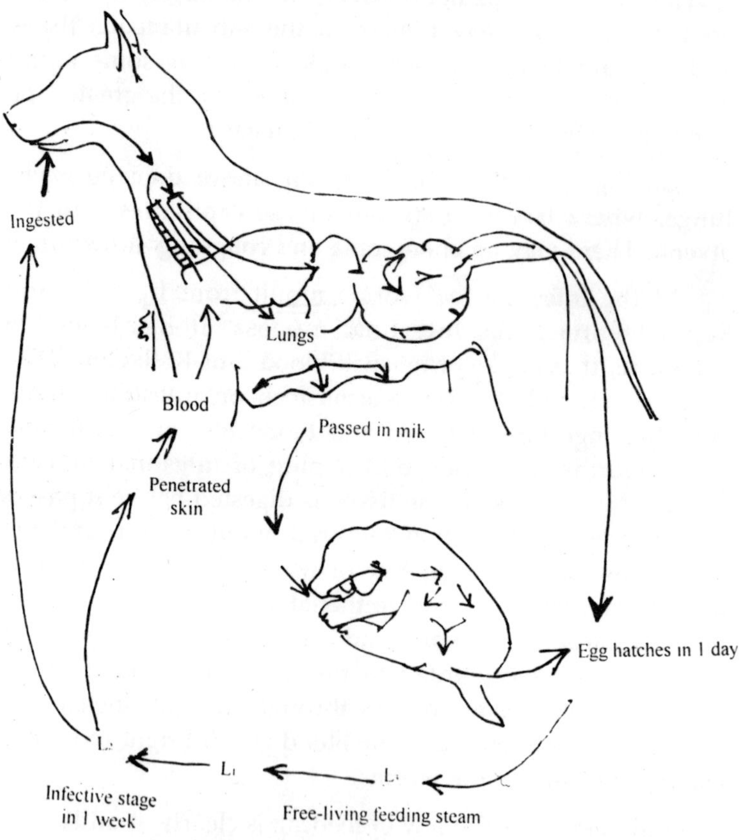

Fig. 4.5 : Life cycle of *Ancylostoma caninum*.

Table 4.1. Some Hookworms of Economic and Medical Importance

Name	*Hosts*	*Geographic Distribution*
Ancylostoma caninum	Canines and other carnivores	Cosmopolitan in warm and warm-temperate climates
A. ceylanicum	Humans, cats, dogs	Asia, South America
A. braziliense	Dogs, cats, and other carnivores	Africa, Central and South America, Gulf Coast of United States
A. duodenale	Humans and some other primates	Subtropical Europe, Asia, South America Africa.
A. tubaeforme	Cats	Cosmopolitan
Nacator americanus	Humans	Tropical: United States, Central and South America, Africa, Orient, Australia
Uncinarla stenocephala	Dogs and other canines	Northern United States and Canada
U. lucasi	Fur seal	Bearing Sea
Bunostomum trigonocephalum	Sheet, goats	Warm, moist climates
B. phlebotomum	Ox	Warm, moist climates
Gaigeria pachyscelis	Sheep	India, Africa

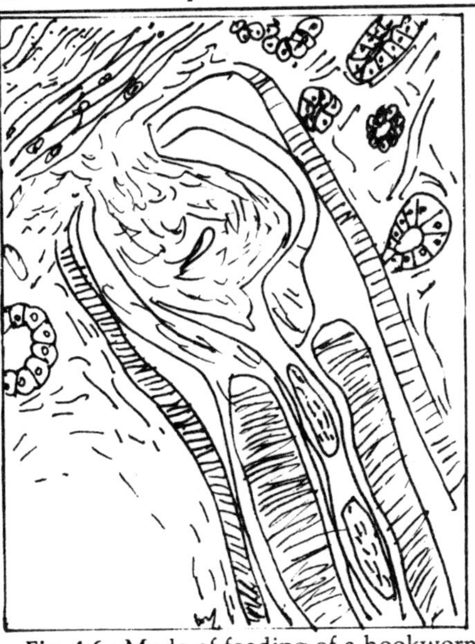

Fig. 4.6 : Mode of feeding of a hookworm.

Whichever view of blood loss is correct, a great deal of blood is passed into the intestinal lumen. Such blood will be partially digested and recycled. The important point is not how much blood leaks from the intestine into the lumen, but rather how much is ultimately lost. The following table shows some aspects of blood loss in *A. caninum* infections.

No. of Worms/LB of Body Weight	Total No. of Worms
8-27	240-934
30-64	371-972
Millilitres of Blood loss/ worm/day	Calculated/ loss/year/ 100 worms
0.07-0.20	2.5-7.3 L
0.01-0.09	0.36-3.3 L

Obviously, the calculated blood loss is not realized, or animals could not survive a moderate infection.

Animals that have been repeatedly infected develop immunity; this observation was used to develop an irradiated larval vaccine against *A. caninum*. A series of studies growing out of the work on lungworm disease in cattle showed that vaccination against hookworms was feasible. A decision was made by Jensen-Salsbery Laboratories to produce and market a hookworm vaccine in the United States. Basic studies done by Miller starting in the 1960s found that it was feasible to vaccinate dogs against hookworm and that they would be well protected against the effects of the worms.

The problems to be overcome for the commercial success of the vaccine were partly biological, partly economic. Miller (1978) reviewed the sequence of events that led to the withdrawal of the vaccine against *A. caninum* after 14 years of basic research and development. A number of hurdles and to be cleared before a product could be made available to veterinarians for use. For example, federal standards for biologics require near sterility, and the fecal culture method of producing L_3s led to so much contamination with bacteria that a sterile culture technique had to be developed. The more laborious procedures for sterile culture of the free-living stages had an advantage in there being less damage to the L_3s at harvest and a longer shelf life for the vaccine. Once the larvae had been irradiated, storage temperature had to be controlled at between 10 and 15°C for a maximum shelf life of 12 to 14 months. Temperatures as low as 4°C and as high as 20°C could be tolerated only briefly. Thus, special shipping containers had to be developed as well as special incubators to be placed in the veterinarian's refrigerator (normally set at 4-6°C).

After the biological problems of production, storage, and adverse vaccine reactions were overcome, the vaccine did not sell well. Part of the problem was a matter of education and part was economics. The vaccine did not produce a sterile immunity. Hookworm eggs appeared in the feces, and it seemed impossible to educate either veterinarians or owners to the fact that low-level production of eggs did not necessarily indicate that the animal was adversely affected. Perhaps of greater importance to veterinarians was the economic return and the fact that adequate control of hookworm disease could be achieved by routine fecal examination and treatment of dogs. Since the market for the vaccine failed to develop, it was withdrawn in 1975 after being in use for about 2 years.

Treatment

A number of drugs are on the market that have a high degree of efficacy, but no treatment should be given until a proper diagnosis has been made. All drugs are toxic to a degree, and the animal may have more than one parasite that requires treatment.

Among the drugs given by mouth are *n*-butyl chloride, toluene, and bephenium. Disophenol can be given by subcutaneous injection. All of these drugs remove 90% or more of mature worms, but do not remove larvae and may vary somewhat in their effect on different hookworm species. Some such as toluene, remove both hookworms and ascarids.

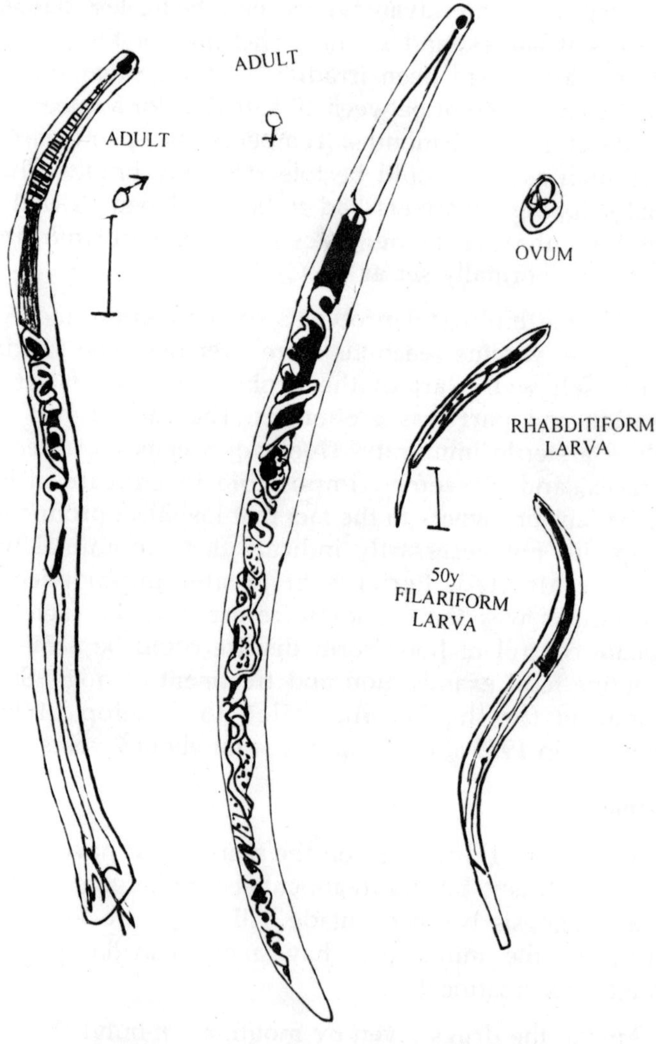

Fig. 4.7 : The hookworm *Ancylostoma duodenale*.

Name of Organism and Disease Caused

Ancylostoma duodenale Crepling, 1845; *Necator americanus* Stiles, 1903, and *A. ceylanicum* cause hookworm disease, hook-worm anemia and ancyclostomiasis.

Hosts and Location

A. duodenale and *N. americanus* are the principal hook-worms that infect humans. Following early migration, they are found in the small intestinal lumen. Hookworms sometimes mature in abnormal hosts, and hosts other than the principal ones can be infected with human hookworms.

Distribution and Importance

Human hookworm is endemic in warm, wet climates, but each species has some limitation on its distribution. Also, in some locations, such as in coal mines in central Europe, the distribution may either be extended or found in small pockets.

Estimates of the prevalence of infection in the human population range from 450 million to 700 million persons worldwide. In his often-quoted article entitled, "This Wormy World," published in 1947, Norman Stoll compiled data indicating that 457 million people were infected world-wide. Thus, there are about the same number of infected people now as 40 years ago. Since that time, the global population has about doubled, so there is some consolation that the per capita prevalence of infection has declined.

Because hookworm infection occurs in the poorest parts of the world, it interacts with other factors to impose a significant drain on the health of people. Inadequate diet and continual blood loss sap the strength and sense of well being. Other helminthic and protozoan infections are common in warm climates, so that any one person is almost certain to harbor at least one pathogenic parasite, and many have more than one.

In the United States, the area of endemic hookworm infection is the 11 southeastern states. From 1910 to 1935 the numbers of persons infected in that area declined from 4.5 million to 1.75 million and has declined still further in the intervening 50 years. However, there are still pockets of relatively high infection in the United States where 12 to 15 per cent of children have been found to be infected. Current data are rather scattered, but there has been a decline in the Southeast from 42 per cent to perhaps 4 per cent in prevalence from early in this century to the present time.

Morphology

In general size, the two principal human hookworms are similar, but can be differentiated statistically:

	A. duodenale	*N. americanus*
Male	8-11 × 0.45 μm	6-9 × 0.3 μm
Spicules	2000 μm	900-1000 μm
Female	10-18 × 0.6 μm	9-11 × 0.35 μm
Eggs	40 × 60 μm	30-40 × 60-76 μm

The two can readily be identified by an examination of the ventral teeth, but differentiation from other closely related species is more difficult. *A. duodenale* needs to be differentiated from *A. braziliense* and *A. ceylanicum,* in approximate descending order of their probability of occurring in humans.

N. americanus is the only member of this genus that has been reported from humans, but there are some similar species such as *N. suillus* from swine that are found in the same geographic areas.

The principal characteristics used in identifying species of hookworms are dorsal and ventral teeth, size and shape of the buccal capsule, dorsal ray of the bursa, length of the spicules and the structure of their tips, the lateral rays of the bursa and the posterior end of the female. Because of the similarity between *A. braziliense* and *A ceylanicum,* the two were not differentiated until Biocca's (1951a) careful study.

Life Cycle

Similar to that of *A. caninum.*

Diagnosis

The principal means of determining whether a person is infected with hookworms is by examining the feces for eggs. The technique most often used is a flotation method, but a sedimentation method can also be used.

In some instances it is important to determine the level of infection in an individual or a community, and quantitative egg counts can provide useful information. Taking a number of studies into account, "light" infections are represented by egg counts up to about 2000-eggs/g of feces. Above 2000 eggs/g infections are considered to be "heavy". There is a direct correlation between the number of worms and the number of eggs per gram of feces; there is also a direct correlation between the number of worms and the severity of anemia. Neither correlation is necessarily linear, but the number of eggs per gram of feces can be used as a guide to possible clinical disease.

The clinical sign of anemia coupled with fatigue are suggestive of hookworm infection in an endemic area.

Host-parasite Interactions

The ways in which hookworm cause damage to their human hosts may be described as follows:

1. Primary, in which the signs and symptoms are directly related to the worms:
   (*a*) Skin penetration

   (*b*) Migration through the lungs and pharyngeal regions

   (*c*) Presence of worms in the small intestine, where they cause direct effects.

2. *Secondary,* in which disturbances of the physiology and/or blood of the host result from chronic infection.

Penetration of the skin causes a slight lesion at the site which may be no more than an itching or prickling sensation; however, upon subsequent infection there is likely to be a more severe reaction due to hypersensitivity.

As the worms break into the alveoli of the lungs and migrate anteriorly, there may be some chest pain and inflammation of the pulmonary and pharyngeal tissues.

When the worms reach the small intestine, diarrhea and abdominal pain occur, among other gastrointestinal signs and symptoms.

The most significant aspect of hookworm infection is anemia as a result of long-standing infection. The severity of hookworm anemia is related most importantly to the number of worms and the nutritional status of the person. In general, anemia results from depletion of iron stores and the inability of the person to maintain a normal amount of haemoglobin.

Removal of blood by human hookworm can be calculated as follows:

	Blood loss	
Species	*Millilitres/ worms/day*	*Millilitres/ 100 worms/year*
N. americanus	0.01-0.04	365-1,460
A. duodenale	0.05-0.3	1,825-10,950

One hundred worms is not generally considered to be a severe infection despite what appears to be dramatic loss of blood. This is because much of the blood is recycled. The greater portion of the protein and iron will be digested, absorbed, and resynthesized into plasma and red cells. The more anterior the site of the worms in the small intestine, the greater the recycling; if the worms lie in the lower part of the small intestine, there will be greater real losses.

An inexorable loss of iron depletes iron stores so that eventually there is insufficient iron to make haemoglobin. The normal amount of hemoglobin (15 g/100 ml of blood) is maintained until iron reserves or stores have been depleted; then there is a decline in hemoglobin to a point where the individual stabilizes at about 50 per cent of normal haemoglobin. A person with such anemia would clearly not feel normally vigorous.

There are other factors contributing to hookworm disease, such as impaired liver function and loss of protein from the gut, but they are beyond the scope of this discussion. Further information may be obtained in Roche and Layrisse (1966) or Miller (1979).

As indicated earlier in the section on *A. caninum*, a high level of protective immunity can be developed in the dog; however, there is no conclusive evidence for protective immunity in humans. A considerable body of circumstantial evidence points to there being a protective immunity. The evidence suggesting a protective immunity is based on the following:

1. Epidemiologic studies
2. Observations after treatment
3. Comparison to infection in other hosts such as dogs.

The epidemiologic pattern of hookworm infection in the human population is an increasing prevalence of infection up to about age 10 years and then a levelling off or decline until age 30 or 40 years; finally, there may be an increase in prevalence in elderly people. One could conclude that the susceptibles are removed from the younger population, leaving the resistant ones, but this seems not to be the case. Following treatment, adults tend to have relatively low worm burdens, but children under 10 years of age reacquire high levels of infection.

The pattern of acquisition and maintenance of worm infection is the same in dogs and people; therefore, humans appear to develop immunity upon repeated or prolonged exposure to infective larvae.

If there is good immunity, the issue of developing a vaccine should be addressed. With the constant drain on the health of most of those 600 million persons infected around the world, a hookworm vaccine would be of great benefit to those countries where the prevalence is high. The development of a vaccine is hampered not by technology, but rather by ethics. There is no way of testing the efficacy and safety of a vaccine for humans in any other host. Developing an irradiated larval vaccine requires that living, perhaps pathogenic, organisms be given to children and then their immunity following challenge observed. Who will assume the risks of such a trial?

The prepatent period in hookworm infections is generally 20 to 50 days, depending on the species, and it generally has been held that once they have infected a host, the worms march inexorably toward sexual maturity. However Schad *et al.* (1973) and Nawalinski and Schad (1974) showed that development might not proceed at the time *A. duodenale* infects a host; sexual maturity can be delayed as long as 40 weeks after infection, nearly six times the usual period. Their work was done in the region of Kolkata, India, where there is a rainy season followed by a long, hot, dry season.

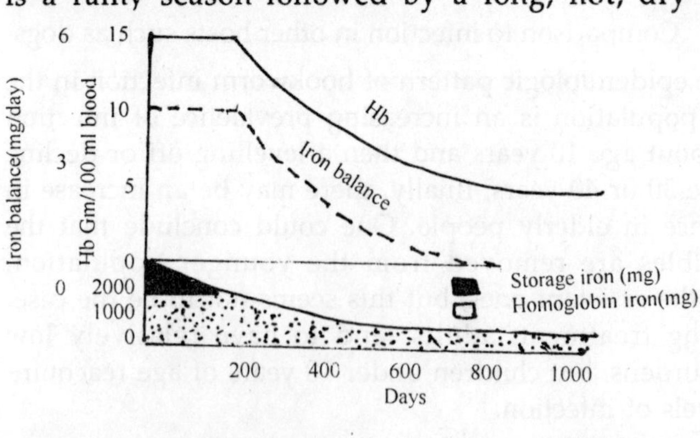

Fig. 4.8 : Iron depletion and the development of hookworm anaemia in humans.

Persons who become infected during the wet season harbor pre-adult worms all through the dry season; egg production commences shortly before the rains begin. It is not known whether the long pre-patent period is inherent in the worms, or whether they somehow sense seasonal change from inside the host and then commence reproduction. Whatever the mechanism, the worms have been selected so that they do not jeopardize their resources until there is a high probability that the life cycle will be completed. The phenomenon of delayed or arrested development has long been known in trichostrongyles of domestic ruminants but this is the first clear-cut instance of arrested development in a nematode parasite of humans.

Disposal of feces deposited on the ground takes place through a combination of weather and microbial and invertebrate animal activity. Invertebrates are the most important in the early dispersal of feces, and in warm climates; beetles seem to be the principal agents that break up and carry them away. In the southern United States, for example, dung beetles (family Scarabeidae) and burying beetles (family Silphidae) can remove the feces of a family of four to six persons during the warm part of the year. It is still not clear whether the beetles are detrimental or beneficial to the development and survival of hookworm free-living stages.

During the middle part of the nineteenth century anemia in coal miners became a problem, or at least came to the attention of the medical community then. Investigation by Italian parasitologists, in the main, showed that people working in mines suffered from hookworm anemia. Working conditions in mines in the nineteenth century left much to be desired, and sanitation was poor. People sometimes worked only partly clothed in cramped, hot, humid, dark mines. If people defecated in the mines, someone was likely to come along later and be exposed to the infective larvae. Thus, special conditions lead to problems with these worms where they would not ordinarily be expected.

Hookworm disease and the prevalence of infection are greatest in the less developed countries of the world, especially those with warm climates. Where there is little sanitary disposal of human waste, subsistence farming, and too little money for the purchase of shoes, hookworms will be prevalent.

A total campaign against hookworm disease involves the following:

1. Mass chemotherapy
2. Sanitary disposal of human waste
3. Wearing shoes
4. Administration of iron salts
5. Ensuring adequate dietary protein.

We tend to think of sanitary disposal of human waste as a panacea for control of intestinal infections, but as Scott and Barlow (1938) showed, nearly complete disposal is needed before there will be significant reduction in transmission. This does not mean that the attempt should not be made.

Mass chemotherapy is most beneficial for those who have had sufficient exposure during their lives to have developed immunity. Children merely reacquire infection after treatment.

Oddly enough, help sometimes comes from unexpected sources. Hookworm infection in Egypt has declined because of the availability of inexpensive plastic shoes. The protection that these shoes provide seems to have reduced the prevalence of, hookworm infection. Merely running an educational campaign on the importance of wearing shoes would probably have had little impact, but the availability of cheap, comfortable shoes has helped to alleviate the problem.

The campaign against hookworm disease in the United State entailed all of the above measures, and was partially successful by the 1930s, prevalence had been reduced by about two-thirds in the southeastern states from what it had been in the early part of the century. The real gains came with general improvement in the economic level during the 1940s, and hookworm disease is no longer a widespread problem.

CUTANIOUS LARVA MIGRANS

Human skin may be invaded by a number of species of nematode larvae, with the result that there are skin lesions at the site of entry. The larvae remain in the skin and cause inflammation, which may last for a few weeks. The pattern is typical of a number of helmith parasites that may invade a foreign host: the normal migration is altered.

The principal hookworms causing cutaneous larva migrans are *A. braziliense. Uncinaria stenocephala,* and *Bunostomum phlebotomum; A caninum* and *A. ceylanicum* play only a minor role. In the rhabditoid nematodes, *Strongyloides papillosus* also causes skin lesions in humans.

In addition to the technical term *cutaneous larva migrans,* the disease is also referred to as *plumber's itch, ground itch,* and *creeping eruption.*

The common names for the disease are descriptive not only of the conditions under which a person may become infected but also of the appearance of the skin. The larvae penetrate the skin, sometimes remaining in place, but sometimes wandering in the skin, causing inflamed, sinuous tracks. Repeated infection may give rise to a more severe reaction at the site of entry indicating the development of hypersensitivity.

Usual treatment is to freeze the skin with a spray of ethylene chloride; the larvae are killed and then cleaned up by phagocytic cells. Thiabendazole ointment is even more effective.

HOOKWORM-PATHOLOGY

Hookworm infection gives rise to a wide range of signs and symptoms indicative of a number of pathological conditions loosely termed 'hookworm disease'. The hallmark of hookworm disease is iron-deficiency anaemia, a condition which should give no cause for concern when nutrition is adequate, or could be readily reversed by iron therapy. Sadly, in hookworm-endemic areas adequate nutrition is the exception, and not only is dietary protein in short supply but it is usually of poor biological value. Similarly, dietary iron supplementation is not readily available, making it difficult for the subjects to maintain the required iron balance. It is for these reasons that iron-deficiency anaemia due to hookworm assumes the importance it does in the tropics. Miller (1979) has given a comprehensive review of hookworm infection. A drastic summary is presented here. For greater details, the reader should consult Miller's review.

Morbidity in hookworm infections can be either acute or chronic: acute infection is due to the activities of migrating larvae and the adults recently reaching the intestine, while morbidity due to chronic infections arise from the activities of established adults in the intestine as well as the consequences of the physiological, biochemical, and haematological disturbances they cause.

Signs and Symptoms of Acute Infection

Penetrating larvae produce inflammation (dermatitis) which is intensified by secondary bacterial infection causing an irritation of varying severity known as *ground itch*. They do not, however, produce the serpentine tracks associated with cutaneous larva migrans characteristic of other non-human hookworm larvae. Although there are no marked pulmonary signs, there are clear signs of upper respiratory tract involvement expressed as laryngitis, pain when speaking and swallowing etc.

On arriving in the intestine the worms induce signs and symptoms including severe abdominal (epigastric) pain, loss of appetite, nausea and vomiting, flatulence, etc. Associated with abdominal pain are indigestion and diarrhoea produced on arrival of the fourth stage larva, and dysentery and constipation arising from the blood-sucking activities of adult worms in the intestine. But the most distinct sign at this stage is anaemia which, when there are adequate iron reserves, need constitute no serious problem. The acute haemorrhage which frequently results in fatal anaemia in canine hookworm infection has only rarely been reported in humans. This condition, 'acute uncompensated haemorrhagic anaemia', to give it its full name, may be more common than is generally thought, particularly in remote and backward areas where high infant mortality is not routinely reported and fully documented.

Signs and Symptoms of Chronic Infection

Signs and symptoms of chronic infection can be classified broadly into two:

1. Primary, which are the direct result of the actions of the worms themselves. These signs recede or are greatly reduced in intensity when the worms are expelled.
2. Secondary, which arise indirectly from the effects of the anaemia caused by the worms. It is possible to remove these signs without expelling the worms, by restoring the haematological level with iron therapy, for example.

Primary Signs of Chronic Infection

These are essentially the same as in acute infection—abdominal pain, diarrhoea, and indigestion, constipation, and a desire to eat soil (geophagy). However, in endemic areas where there is continuous reinfection, these signs may be due to new infection.

Secondary Morbidity in Chronic Infection

The main condition here is iron-deficiency anaemia which develops slowly and insidiously and may not become apparent for years, especially with light infections and adequate iron intake or iron reserves. Of particular significance, the condition can be effectively managed, corrected, or prevented altogether by suitable dietary intervention, especially by the administration of supplementary iron. Foy and Nelson (1963), Gilles, Waston Williams, and Ball (1964), and Roche and Layrisse (1966), have given good accounts or iron-deficiency anaemia. Unfortunately, the following complicating inter-current conditions have been known to hamper the effectiveness of iron therapy:

1. Dietary deficiency
2. Interference with iron absorption
3. Kwashiorkor
4. Chronic bacterial infections like tuberculosis and syphilis
5. Parasitic infections like amoebiasis, malaria, and schistosomiasis.

In addition to iron-deficiency anaemia in chronic hookworm infection, the following conditions also develop: hypoproteinaemia, hypoalbuminaemia, and oedema. Other secondary signs and symptoms include cardiac insufficiency and congestive cardiac failuie, weakness, breathlessness, palpitation, systolic murmur, angina pectoris, vertigo, fainting, etc.

However, just as serious as the effects of anaemia and other conditions are the deficiencies apparent in the health profiles of individuals and communities ostensibly induced by hookworm disease. Thus:

1. Productivity is greatly undermined because of disinclination and inability to work.
2. Hookworm subjects tend to present an image of dullness, stupidity, confusion, poor memory, etc.
3. The effects on children are more severe. Chronic infection has been held responsible for the stunting of the physical and mental development of children;

indeed, such children have been reported to be the most stupid in their class in endemic areas.

4. Hookworm disease has been incriminated in many reproductive disorders including;

   (*a*) Delayed puberty.

   (*b*) Miscarriages, abortion, stillbirths.

   (*c*) Up to 70 per cent of foetal, neonatal, and early infant mortality, as well as 25 per cent of maternal mortality in endemic areas have been attributed to hookworm disease.

Anaemia due to hookworm is one of the commonest causes of hospitalization in tropical Africa and is one of the principal causes of heart disease. However, the severity of the anaemia depends on the worm load and the dietary intake of the patient. For example, in Nigeria, it was estimated that 800 worms (with egg counts of 20,000 ova per gram of faeces) produced significant clinical anaemia, while in East Africa and Gambia, smaller worm burdens were responsible. This disparity is apparently due to the higher iron content of the Nigerian diet estimated at 40 mg per day. Of this amount, however, only about 5 per cent of the iron is absorbed and since normal daily requirement of iron is in the range of 0.8 to 3.2 mg, any reduction of the absorption below 5 per cent could upset the iron balance adversely. Also, the degree of morbidity depends on the species of hookworm and this is probably related to the amount of blood loss caused. Thus *A. duodenale* infections (with 0.15 ml per worm per day of blood) are more severe than *N. americanus* infections responsible for only. 0.03 ml per worm per day of blood loss.

One study in Northern Nigeria must be pointed out. Among the Fulanis where iron-deficiency anaemia is rampant, Oomen (1975) showed that the mild degree of hookworm infections means that hookworm cannot be held responsible as a major contributory factor.

Mortality

Mortality due to hookworm disease has not been widely reported. However, Bwimbo (1970) reported that hookworm anaemia is the third most frequent cause of death in 1 to 5 year old children in Uganda, while Trowell (1956) and Fowler, Knight, and Patel (1968) consider that death from cardiovascular inadequacy and circulatory collapse brought about by hookworm disease is not uncommon. Finally, Miller (1979) is convinced that the incidence of hookworm mortality may be higher than generally accepted to be the case. This is particularly so in cases of infant and child mortality in many endemic areas which are neither reported, investigated, nor documented.

Pathophysiology

A comprehensive review has been given by Roche and Layrisse (1966). Pathophysiology is presented in the form of blood loss (enterorragia) accompanied by iron loss. Blood is lost;

1. When worms suck blood required for their own nutrition and respiration. It appears that they suck more blood than is actually necessary to satisfy their needs.
2. When blood is spilled around the worm's head.
3. When blood leaks from damaged blood vessels in the intestinal lining.

The amount of blood loss is estimated by using the technique of radioactive labelling of circulating erythrocytes with Cr. With this technique, it has been estimated that up to 200 ml per day may be lost by patients with heavy infection. The amount of iron lost is also important. This, however, depends on the amount of blood lost as well as the haemoglobin values at the time bleeding occurred. About 40 per cent of the iron in the blood lost may be reabsorbed although confirmatory proof is required for most reports.

Another pathological feature is haemolysis (erythrocyte destruction) caused by haemolysins produced by the hookworms.

On the whole, therefore, iron-deficiency (hypochromic) anaemia is the most important factor in the morbidity and mortality of chronic hookworm disease. This condition can be reversed by iron therapy and, in any case, should have little chance of developing in the first place if there is an adequate iron-rich diet or if supplementary dietary iron is introduced into the regular diet to restore the iron balance. As it happens, hookworm endemic areas tend to coincide with areas where iron deficiency is rampart, so the people ordinarily have a hard time achieving the minimum iron and blood levels, let alone having to contend with heavy hookworm burdens. Miller (1979) has given an excellent summary of iron balance in relation to hookworm infection.

HOOKWORM—DIAGNOSIS AND TREATMENT

The specific diagnosis is finding the characteristic eggs in stool samples. It is often necessary to estimate the output of eggs as a measure of this worm burden, and techniques for this are described by Mardsen (1977). It is usually not possible to distinguish between eggs of *N. americanus* and *A. duodenale* and so they are generally reported as hookworm eggs.

A sensitive method for hookworm identification is the 'test-tube filter-paper culture method'. Here, faeces is smeared at one end of a strip of filter paper, while the other end is dipped into water in a test-tube. The larvae which subsequently accumulate in the water can be identified and estimated. This method not only makes it possible to separate larvae of hookworm and *Strongyloides,* but to distinguish between the two species of hookworm as well.

Intradermal tests have been developed using various hookworm antigens to produce hypersensitive reactions. This is useful for mass screening for hookworms in endemic areas. The general appearance of a person suffering from hypochromic anaemia in a tropical area should serve as a clue.

Light infections require no treatment beyond iron therapy and indeed severe symptoms may not even develop if an iron rich diet is available. In heavy infections, however, supplementary dietary iron will alleviate the symptoms—if not reverse them altogether. Chemotherapeutic agents include tetrachlorethylene, which is an effective and cheap hookworm expeller, particularly good for mass treatment. However, it causes *Ascaris lumbricoides* to not together causing intestinal blockage with serious consequences. Bephenium hydroxynaphthoate (Alcopar) and mebendazole are preferable in cases of concurrent infections of hookworms and roundworms, being effective against the two.

There is no conclusive evidence that man can develop a functional protective immunity to hookworm infection. Miller (1979) is convinced that more research in this field may reveal immunity similar to that operating in the dog-hook-worm relationship.

HOOKWORM—EPIDEMIOLOGY

The epidemiology is influenced by the standard of sanitation and the existence of conducive environmental conditions for the development of the eggs and survival of the larvae. The features which promote transmission, as well as those which are unfavourable, are well recognized. The following is a summary in outline of the principal factors gleaned mainly from the succinct review of Schmidt and Roberts (1981).

Chief among the factors necessary for transmission to occur are:

1. Contamination of the soil through direct defaecation or deposition of faeces on the ground; in other words, poor sanitation.
2. Exposure of skin to infected soil, *e.g.*, by going about bare-footed.

3. Suitable environmental conditions including:
   (*a*) Eggs require warmth, shade, and moisture for development.
   (*b*) Larva require the following:
      (*i*) optimum temperature 23-30°C,
      (*ii*) oxygen, because their metabolism is aerobic, and
      (*iii*) pH near neutrality.
4. Loose, humus soil with reasonable drainage and aeration.
5. Annual rainfall of 90 cm is necessary but must be spread uniformly throughout the year. Nwosu and Anya (1980) have investigated the effect of rainfall in Nigeria.

The following factors are unfavourable:

1. *Dryness.* Larvae are sensitive to desiccation, although they can survive on the film of water surrounding soil particles in otherwise dry soil.
2. *Low temperatures.* Less than 17°C will inhibit the development of eggs and frost is lethal to eggs and larvae.
3. Presence of undiluted faeces inhibits development and urine is lethal.
4. Direct sunlight.
5. Acid or alkaline soils.
6. Strong chemicals are lethal to larvae.
7. Salt in the soil inhibits hatching and is fatal to larvae.
8. Heavy clay, coarse, sandy soils and water-logged soil.
9. Alternate drying and moistening are damaging to the larvae, *e.g.* rainfall followed by long dry spells is unfavourable.

Other factors of epidemiological significance include:

1. Poverty and ignorance.
2. Shady locations as found in coffee, banana, and sugarcane plantations expose workers in such areas to greater risk of infection.
3. On the other hand, workers in rice paddies are free of infection because of flooding.
4. People who prefer defaecation sites which are usually cool and shaded are exposed to continual reinfection when they repeatedly visit these sites.
5. Race, sex, and age are all factors of epidemiological importance. White people are ten times more susceptible than blacks and children more than adults and women more than men.
6. The nutritional status and iron reserve status of the population is very important.
7. Longevity of larvae; most can survive and remain infective for up to three weeks under ideal environmental conditions and even up to one year in protected areas like mines.

Nwosu (1981) has given an interesting account of the community ecology of soil-transmitted nematode infections of humans in Nigeria.

Distribution of Hookworms in Tropical Africa

Human hookworm infection is widespread n tropical Africa, most surveys showing prevalences of 20 to 50 per cent although quite a few exceed 50 per cent. With prevalences of 63 to 100 per cent among the general population and 90 to 97 per cent among school children in all nine countries. Liberia must rank amongst areas of highest hookworm endemicity. From all accounts, both species of hookworm seem to occur in tropical Africa, although *N. americanus* is the commoner of the two. Thus, for example, in Kampala, Uganda, 99 per cent of people infected had

Necator while only 4 per cent had *Ancylostoma* and a similar trend was observed in Liberia. Furthermore, while *N. americanus* is ubiquitous, *A. duodenale* tends to be restricted to coastal areas in both Kenya and Liberia. Also, while infections with only *N. americanus* is the rule, *A duodenale* in general tends to occur in mixed infections with *N. americanus.* In most surveys, however, no attempt has been made to distinguish between the two types of infection. Udonsi, Nwosu, and Anya (1980) have investigated the population structure of *N. americanus* larvae in contaminated farmlands in Nigeria.

HOOKWORM—PREVENTION AND CONTROL

This will depend on improvement in standards of sanitation raising the nutritional status of the population especially in relation to iron content, and mass treatment of the community with a suitable worm expeller (vermifuge) periodically, say two or three times a year, to keep the worm load within limits. These measures tend to reinforce the observation that prevalence of hookworm infection is related to social status, the poorer sections of the community being more liable to infection. Admittedly, the more affluent sections of the community have access to an adequate diet to compensate for the effects of the worms, if indeed they do get infected, but the higher incidence of iron-deficiency anaemia among the lower social groups in tropical areas is not only the result of mal-and undernourishment but because 'sanitary habits and facilities' are worse, footwear less commonly used and hence the chances of hookworm infection are higher and anaemia more liable to develop.

Chapter—5

Trichinella Spiralis

Trichinella spiralis infection is one helminthic infection of man that is more prevalent in the advanced societies of Europe and the United States than anywhere else. It is endemic throughout the Arctic, but up to comparatively recently it was unknown in tropical Africa and other tropical areas of the world. Steele (1976) and Steele, Arambulo, and Moon (1976) have reviewed the worldwide distribution of the Parasite. Ever since Forrester, Nelson, and Sander (1961) reported the isolated outbreak of trichinosis in Kenya as the first record of human trichinosis in Africa south of the Sahara, most authorities have accepted that the disease must be widespread in tropical Africa. In spite of this widely held belief only a few cases of human trichinosis have up till now been reported from tropical Africa. In fact, it is fair to say that the information to go on derives from reports of isolated and sporadic outbreaks—mainly, one in Kenya, one in Tanzania, and one from Dakar, Senegal. Some of the reasons adduced for this include: poor diagnostic facilities, difficulty of diagnosis, and possibility of overlooking or misdiagnosing the disease because its symptoms are similar to those of other tropical infections—which is all very well. But medical services have improved and service which has been ablé to diagnose such a wide array of parasitic infections throughout the continent could not have failed to throw up trichinosis if it were anything but the disease of extreme rarity that it is. The incidence of trichinosis in wild animals in tropical

Africa is, however, another matter because high prevalences have been reported in surveys in Kenya by Nelson, Guggisberg, and Mukundi (1963), in Tanzania by Sachs (1970), and in Senegal by Gretillat (1970), Gretillat and Chevalier (1970), and Gretillat and Vassiliades (1967).

Trichinella is the only genus of the family Trichinellidae of the order Trichurata, and for a long time it was considered to be monospecific, consisting of the single species *T. spiralis*. Many strains or biological races of this species have been recognized, differing in their degree of infectivity in various experimental hosts, their pathogenicity, and in other physiological features, though they are morphologically indistinguishable. Many author recognize these merely as strains, while others have elevated them to specific status: *T. spiralis* for the north temperate 'strain', *T. nelsoni* for the tropical, and *T. nativa* for the arctic 'strains', and even a fourth species *T. pseudospiralis*. The controversy on the speciation of *Trichinella* is very interesting, but beyond the scope of this book. Sukhdeo and Meerovitch (1977) have presented fresh experimental evidence which tends to support the existence of three species, suggesting that they

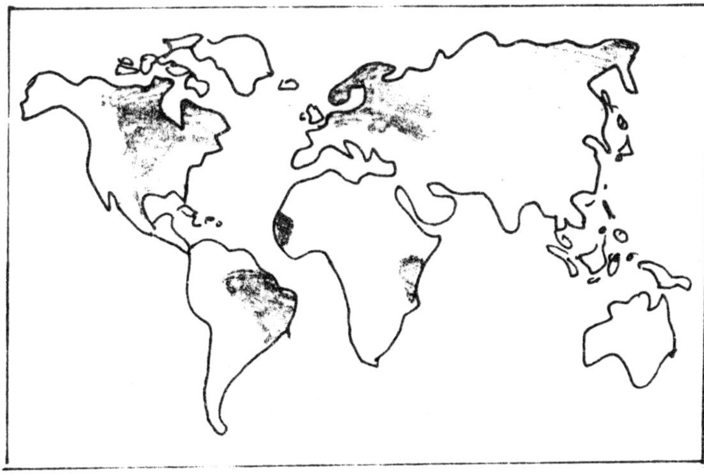

Fig. 5.1 : Map of the world showing the distribution of *Trichinella spiralis*.

are 'incipient species' on the way to becoming completely separated as a result of 'their geographical and hostal isolation'. Madsen (1976), on the other hand, in a most interesting review shows that there is insufficient scientific evidence to justify the creation of new species whose descriptions 'do not in any respect stand up to the international code'.

Morphology

1. *Trichinella spiralis* shows most of the features of the order Trichurata—stichocytes, lack of spicules in males, single female gonad, terminal anus, etc.
2. Size: it is the smallest human nematode parasite, males measuring 1.4 to 1.6 mm and females 3.0 to 4.0 mm in length.
3. The whip-like appearance characteristic of the order is not so marked, the body merely tapering towards the anterior end.
4. Male has two large papillae.
5. Females are oviviparous: the posterior part of the uterus is filled with developing eggs, while the anterior portion contains hatched larvae.

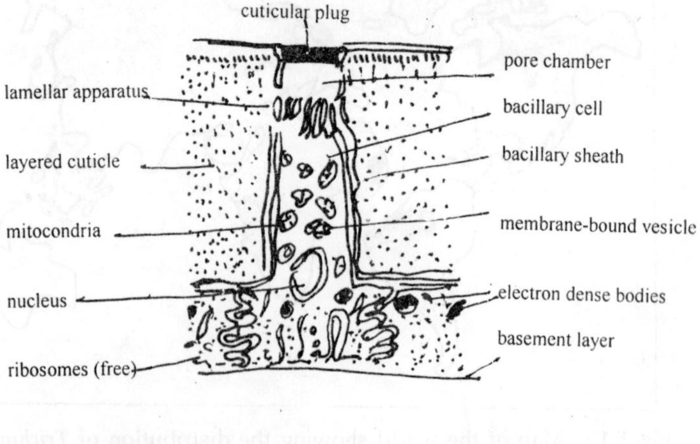

Fig. 5.2 : *Trichuris suis*—ultrastructure of bacillary cell.

Life Cycle

There is one remarkable feature of the life cycle of *T. spiralis;* the larvae and adults develop in different organs of one animal, *i.e.,* a single animal serves as both intermediate and definitive host. The adults live in the intestine of man and other suitable hosts. After copulation, the females burrow into the mucosa and submucosa where, being oviviparous, they produce live larvae (about 1500 per female). These larvae then penetrate further, finally reaching the blood stream which distributes them to various organs of the body. They tend to settle, however, in skeletal muscles, especially muscles of the diaphragm, jaws, tongue, etc. Here they penetrate individual muscle fibres and grow up to ten times their original size, reaching up to 1.0 mm in eight weeks. During this period they begin to coil, forming the characteristic spiral. Host tissue reactions form a lemon-shaped cyst around the larva which is now known as the *encapsulated larva.* Within this cyst, the larva can survive for several months or years and still be viable. However, after about 6 to 18 months, calcification of the cyst and later of the larvae themselves sets in. Infection occurs when infected muscle is consumed by a suitable host. The larvae are released in the duodenum and develop rapidly, reaching sexual maturity in only 40 hours. Shanta and Meerovitch (1967*a*, 1967*b*) have given good accounts of the intestinal and muscle phases of the life cycle.

Hosts of *T. spiralis*

More than a hundred species of mammals have been reported as host of *T. spiralis.* In the temperate region, there are two well-defined cycles: the *urban cycle* which is maintained in rats, domestic pigs, and man and the *sylvatic cycle* in which wild carnivores like bears, foxes, badgers, weasels, etc., and wild boar are involved. In the Arctic region, the cycle is maintained in polar bears, arctic foxes, wolves, mink, dogs, walruses, and seals. In tropical Africa, surveys of animal hosts have only been carried out in Kenya by Nelson, Guggisberg, and Mukund (1963), in Tanzania by

Sachs (1970), and in Senegal by, Gretillat (1970), Gretillat and Chevalier (1970), and Gretillat and Vassiliades (1967). The results of these surveys have been summarized by Nelson (1972) and reveal that the sylvatic cycle is the dominant cycle, being maintained in large canibalistic and carrion-feeding carnivores, especially spotted and striped hyaenas, leopards, jackals, serval, lions, as well as wild suidae like bush pigs and warthogs. However, domesticated pigs and domiciliated and wild rodents as well as numerous viverids and mustelids

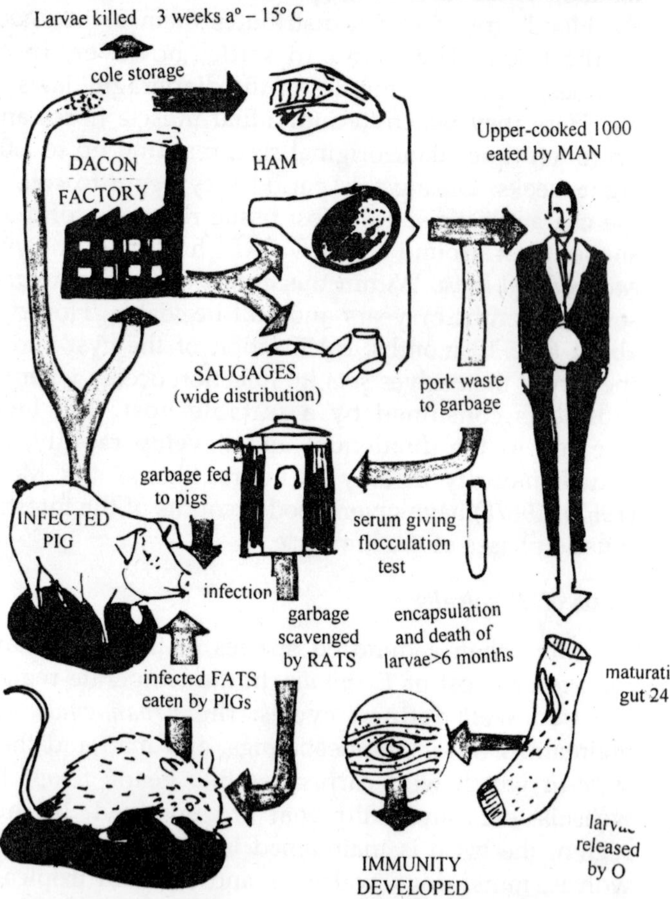

Fig. 5.3 : *Trichinella spiralis*—life cycle and probable mode of transmission in Europe and American communities.

which feed mainly on rodents have not been found to be infected. In this connection, a Kenyan strains of *T. spiralis* has been recognized by Nelson and Mukundi (1963) which is of low infectivity to rats and domestic pigs. Similarly, a West African strain has been recognized by Gretillat and collaborators in Senegal, which also has low infectivity in rodents and domestic pigs, although guinea pigs appear to be more susceptible to it than to the Kenyan strain. In an extensive series of experiments on the susceptibility of various domestic and wild animals to the West African strain, Gretillat and Vassiliades (1968*b*) found that the domestic cat was the most highly susceptible, closely followed by the domestic dog. The domestic pig had very low susceptibility, while rats, mice, Gambian rats, hedgehogs, and rabbits were refractive to infection. The primates, *Papio papio* and *Eurythocebus patas,* were very susceptible while the warthog, which is regarded as an occasional host, was difficult to infect experimentally. Finally, carnivores proved to be the preferred hosts in West Africa. There is apparently an 'intestinal barrier' in the pig which prevents most larvae from developing to adulthood, while a 'muscular barrier' causes rapid calcification of 60 to 100 per cent of the larval cysts. Finally, and most importantly, it has been established that serial passage of this strain from wild carnivores through white rats increased its infectivity and rendered it capable of infecting domestic pigs. This is of great epidemiological significance because pigs can readily become susceptible when exposed to larvae which have undergone repeated passage, a fact stressed by Madsen (1976) in connection with the validity of strains.

Pathogenesis and Pathology

The symptoms of trichinosis are characterized by vagueness, and mimic symptoms of other conditions such that over fifty diseases have been diagnosed for it. Details of the pathogenesis and pathology in tropical Africa have been given in the case reports of the outbreaks by Bura and Willett (1977) and Forrester, Nelson, and Sander (1961).

According to Kangan (1978), of some seventy-five symptoms and signs associated with trichinosis, the most important in order of frequency of occurrence are: fever, gastro-intestinal symptoms including diarrhoea, muscle pain, nausea, oedema of the face and eyelids, eosinophilia, rash, weakness, and neurological symptoms. Death, when it occurs, is the result of heart failure, respiratory complications, toxemia, or kidney malfunction. All these symptoms are the results of the effects of adult worms in the intestinal mucosa, larval migration, and their penetration and encapsulation in the muscle fibres.

One important feature of the East Africans cases is the density of larvae per gram of muscle tissue which is the highest ever recorded for man. In Kenya, larval density ranged from 1850 to 5190 while in Tanzania up to 6000 cysts per gram was reported, with an average of 3390. These figures exceed, by a wide margin, the findings in Liverpool where densities of 7.5 and 119 larvae per gram were classified as severe or very severe respectively. The fact that these densities produced conditions of equal clinical severity suggests that the East African strain is of higher infectivity and lower-pathogenicity than the strain in Liverpool.

Diagnosis and Treatment

Diagnosis of trichinosis presents a number of difficulties:

1. Eggs or larvae are usually not detected in faecal samples.
2. Light and moderate infections show no symptoms.
3. Even heavy infections present vague symptoms that are usually confused with other conditions. Thus trichinosis may be overlooked or misdiagnosed if it is not suspected in the first place.

There are a number of diagnostic techniques for detecting encapsulated larvae in the muscles:

1. Digesting a biopsy specimen of muscle in artificial gastric enzymes such as pepsin and dilute hydrochloric acid overnight, and recovering the larvae by sedimentation in a tube.

2. Using a *trichinoscope* to compress a muscle sample between glass plates, such that it appears semi-transparent under the light microscope, revealing the larvae.

Unfortunately, larvae in muscle biopsy specimens in cases of early and light infections are not easy to find. Furthermore, selection of sites for diagnostic biopsy depends on knowledge of predilection sites of larvae in the muscle of the body, but not many detailed studied of this have been reported. However, it is generally assumed that, as in rats and pigs, the diaphram is the preferred site followed by the jaws, tongue, larynx, and eye. Biopsy specimens are however obtained from the gastroeneminus or deltoid muscle.

Other tests include:

3. Immunodiagnostic techniques: flocculation tests and intradermal tests, although positive reaction with the latter does not necessarily mean the existence of current infection.
4. Xenodiagnosis, in which muscle biopsies are fed to laboratory rats and the presence of infection is confirmed if larvae develop in the muscles of the rat.
5. The presence of peri-orbital oedema, especially when it occurs in a group of patients, is a useful diagnostic clue.

There is no satisfactory drug for specific treatment of trichinosis. Thiabendazole has been shown to kill encapsulated larvae and adult stages in experimental mice, but results of clinical tests with man are equivocal. At present, efforts are directed towards relieving the symptoms, and for this corticosteroids and analgesics come in handy. A review of chemotherapy of trichinosis has been given by Campbell and Blair (1974).

EPIDEMIOLOGY AND CONTROL

The literature on the epidemiology of trichinosis in the temperate zones of Europe and North America is extensive. As stated above, there are two maintenance cycles: the sylvatic cycle involving wild carnivores and their prey or carrion and the urban or domestic cycle involving the 'man-domestic pig-rat triangle'. The situation is graphically represented by Nelson (1972). In this region, the urban cycle plays the dominant role in epidemiology, man becoming

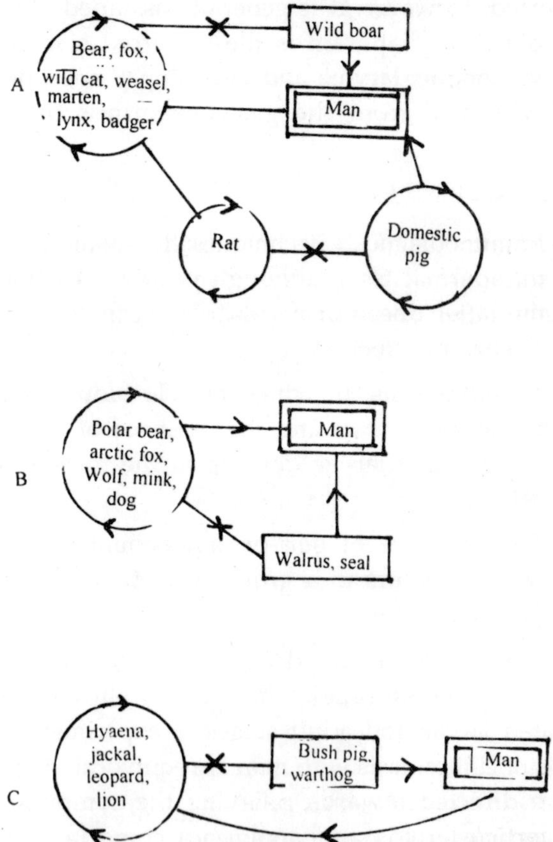

Fig. 5.4 : The maintenance of trichinosis in various regions of the world: A, temperate region, B, arctic region. C, Tropical Africa.

infected by consuming undercooked pork and infected pork products like sausages. Rats become infected by practising cannibalism or by feeding on rat faeces containing infective larvae. Pigs themselves maintain the cycle by being fed garbage containing infected pork scrap and offal, or by eating each others tails, especially in overcrowded piggeries. It is also believed that pigs can gain infection by eating rats in their sties. According to Madsen (1976) this 'predation theory' is responsible for the lack of proper appreciation of the significant role of the encapsulated larva. The evidence has been piling up since the last century that the encapsulated larva is in effect a free-living stage, surviving in the muscle fibres of the decomposing carcass (carrion), a situation analogous to the free-living eggs or larvae of other nematodes. This evidence, according to Madsen, has been largely ignored because of the dominance of the predation theory. A fresh look at the domestic cycle will indicate that what happens is that pigs gain infection by eating dead and decomposing carcasses of rats, while rats gain access to decomposing scraps of infected pork.

In tropical Africa, the domestic cycle is of little or no significance, mainly because, as stated above, the domestic pig and rats have low susceptibility to the African strain of the parasite. In this respect, the epidemiology is remarkably similar to that of the Arctic region. In tropical Africa, therefore, the sylvatic (=feral) cycle is the dominant one, the cycle being maintained in carnivores. In East Africa, it is significant that the hyaena is the commonest host with a 44 per cent infection rate. It is significant because the hyaena is a scavenger, thriving mostly on carrion and occasionally practising cannibalism as well. According to Madsen (1976), wild carnivores consume more carrion than is generally realized, thereby reinforcing the belief in the importance of putrefaction of infected meat in the epidemiology of trichinosis. With the carnivores feeding on carrion or on flesh of other carnivores and even practising cannibalism, it can be seen that wild rodents are not essential for the cycle in Africa. However, wild suids like bushpigs and warthogs come into the cycle again, either through carrion-feeding or cannibalism.

Since people in tropical Africa, as in most other parts of the world, do not relish the flesh of carnivores, transmission is most likely to occur by eating the undercooked flesh of wild suids. Indeed, the three well-documented cases from Kenya, Tanzania, and Senegal involved the consumption of the flesh of the bushpig and warthog. The Senegal case is interesting because it involved nine Europeans in Dakar who indulged themselves by feasting on barbecued warthog, an obvious act of indiscretion in the tropics, or anywhere else for that matter !

With regard to the outbreak in Kenya reported by Forrester, Nelson, and Sander (1961), Nelson (1972) attributes it to change in human behaviour brought about by the Mau Mau 'insurrection' which made young Kenyans abandon old Kikuyu tradition of disdain for the flesh of wild animals, and eat undercooked meat to bushpig. However, this explanation may not be wholly acceptable because, in the first place, people do not require an insurrection (a struggle for political independence in this case) in order to change or discard traditional practices. In the second place, wild pig is commonly consumed in different parts of Africa where trichinosis is not a serious problem.

The absence of trichinosis from most parts of tropical Africa has intrigued most workers in the industrialized countries of the temperate zones. In their search for a scientific explanation of this phenomenon, a variety of opinions have been expressed. For example, Nelson (1972), Nelson, Guggisberg, and Mukundi (1963), and Pawlowski (1974) believe that: *T. spiralis* infection is widespread and will come to light if we search hard enough. Another view is that trichinosis has been missed so far simply because of its poor diagnostic features or because it presents symptoms which simulate other more common and familiar tropical diseases, so that the suggestive symptoms are overlooked if trichinosis is not suspected in the first place. Finally, it has been predicted that there is the possibility of changing epidemiology either through the introduction of infective strains with the importation of exotic breeds of pigs in the expanding pig industry or as a result of 'repetitious transmission' through rats and pigs which will increase the infectivity of the strain. However, a simple explanation may be that this is an infection which is endemic in the temperate region but sporadic in tropical Africa.

In conclusion, the essential feature of the epidemiology of trichinosis in tropical Africa is that sylvatic trichinosis is prevalent and the cycle is maintained in wild, cannibalistic, or carrion-feeding carnivores and wild suids. Man, however, occasionally become involved by eating undercooked flesh of wild suids, and this results in outbreaks of the disease in isolated part of the continent.

Prevention and Control

The obvious preventive measure is by eating well-cooked pork. Control measures include cooking port at a temperature of 55°C for 15 minutes per pound or alternatively freezing pork products at –40°C for 30 minutes, –25°C for 24 hours, or at –15°C for 6 days (Smyth, 1976). Such treatments are adequate to kill the encapsulated larvae. Also, drying or salting pork for long periods renders it safe for human consumption.

It is now generally accepted that the prevalence of trichinosis is declining throughout the world; indeed, it was never of great importance in the first place. Reduction in transmission in the United States was achieved as a result of legislation against feeding undercooked or raw garbage to pigs. However, the irony, according to Nelson (1972), was that this legislation was not directed primarily against the human disease, but rather against *Vesicular exanthema,* a virus infection of swine of great veterinary and economic importance to the piggery industry !

By the forty-third day, Onabamiro found that the numbers of males and females were in a 1 to 1 ratio and although already sexually differentiated, the males had not yet developed spicules and the females were unfertilized. Copulation takes place at about 3½ months, after which the males die and become encysted and the females migrate down towards the extremities. They emerge and start to discharge larvae 10 to 14 months after infection.

GUINEAWORM—PATHOGENICITY

The first signs of the disease appear when the gravid female is about to emerge from the subcutaneous tissues. It then provokes the formation of a small blister accompanied by itching and intense burning pain, together with allergic reactions in the form of generalized urticaria, fever, giddiness, gastro-intestinal symptoms, and dyspnoea. After a few days, the blister bursts. The itching sensation subsides when the blister is immersed in cold water and it is then that the anterior part of the female worm is extruded and larvae discharged. The release of larvae takes place periodically till all the larvae are discharged and the entire worm expelled. The whole process lasts from 2 to 4 weeks and the sore heals rapidly after that. The effect of the disease when it takes this course is mild, with minimum or no discomfort, and according to Kale (1977) accounts for 57 per cent of the cases in Nigeria.

Muller (1971) gives a detailed account of the pathogenicity of the disease and shows that there are complications; it is these that results in incapacitation of the patient. Without going into too much detail, tissue reaction as a result of the release of larvae produces abscesses which can lead to chronic ulcerations. When a worm bursts deep in the subcutaneous tissues, it produces an abscess which can become secondarily infected by bacteria. Secondary bacterial infection can also occur along the track of the worm, especially if it is damaged within the tissues or if it retracts into the body. These conditions give rise to serious disabilities, and when there is involvement of the joints of the leg produce such conditions as fibrous ankylosis, contractures of tendons, septic arthritis, and aseptic arthritis. Cases of guineaworm abscesses in extradural space of the spinal canal resulting in paraplegia and motor weakness of the legs have been reported. Sometimes, worms which do not emerge die and become calcified near joints like the pelvis, knee joints, etc. It is possible that such worms contribute to arthritis observed in such patients. Worms have also been found in the urino-genital system, and St. George (1975) reported cases of retroplacental situation of the worms causing bleeding during pregnancy and, in one case, abortion. Finally, guineaworm ulcer is one of the commonest sites for entry of tetanus spores, so the risk of tetanus as a complication of dracontiasis is very high.

On the whole, although about 30 per cent of infected cases end in periodic incapacitation and prostration, causing great suffering, only about 0.5 per cent end up in permanent disability, while the mortality rate is only 0.1 per cent. In Nigeria, over 90 per cent of the blister/ulcers marking the sites of emergence of the worms are located in the lower limbs, particularly the ankles and feet, while the trunk and abdomen account for 2 to 6 per cent, the upper limbs 1.8 to 3.5 per cent, and the head and neck 0.2 to 1.0 per cent. Some of the unusual sites include the scrotum, vulva, penis, buttocks, and chin. Wurapa, Belcher, and Ward (1975) has given a clinical picture of the disease in Ghana.

GUINEAWORM—DIAGNOSIS AND TREATMENT

Diagnosis

In an endemic area, dracontiasis is not difficult to recognize, particularly must before the female worm begins to emerge. Then the patient experience the intense itching and burning pain at the site of the blister characteristic of the disease. This is also accompanied by general urticaria. When the blister bursts, specific diagnosis can be made by finding actively moving larvae when a sample obtained from the milky fluid, to which a drop of cold water has been added, is examined under the microscope. Most of the immunological diagnostic tests so far developed have little practical application, and in any case are effective only for patients with already patent infections. In a bid to diagnose the disease early in its long pre-patent period, Muller (1970b) has developed the indirect fluorescent antibody technique using sections of adults or first stage larvae as antigens. With this method, antibodies have been detected in experimentally infected rhesus monkeys 4 to 8 months before the emergence of the worm, a valuable technique if it could be made to work in cases of human infection.

Treatment

The time-honoured method of winding the emergent female worm around a small stick is still widely practised. Care must be taken not to break the worm. The process of extraction is enhanced by immersing the ulcer in ice cold water and winding out a few centimetres a day, the whole process being completed in about 3 weeks. It is essential to prevent secondary bacterial infection of the ulcer by applying sterile dressings, etc.

So far, no drug has been shown to have direct anthelminthic action on the worm although niridazole, thiabendazole and metronidazole have been shown to reduce host tissue reaction and facilitate the removal of the female worms without pain. However, mass treatment with these drugs is at present not feasible because they are too toxic and expensive. In any case, none of these drugs has a prophylactic effect, so infection can always recur after successful treatment.

Where modern medicine seems to have failed to provide a remedy, traditional African medicines have been widely acknowledged to provide varying degrees of cure. The Director of the Ghana Centre for Scientific Research into Plant Medicine provided a well-documented report of how a decoction of the root of *Combretum mucronatum* (long prescribed by traditional herbalists) procured a 97.7 per cent cure rate with complete extrusion of the worms, reduction in inflammation around the ulcers, and healing of the ulcers in two weeks following the application of sterile palm oil. Similarly, *Mitragyna stipulosa* also gave a 52.2 per cent cure rate. Finally, he also proved that leaves of *Elaeophorbia drupifera* and *Hilleria latifolia* taken in palm soup are effective against guineaworm infection.

GUINEAWORM—EPIDEMIOLOGY

The epidemiology of dracontiasis in tropical Africa presents many interesting features which for the sake of brevity can only be outlined here. For more information the reader is referred to Belcher *et al.* (1975). Kale (1977), Lyons (1972), Muller (1971, 1979), Onabamiro (1951, 1952c), Scott (1960), and Ward *et al.* (1979).

1. Dracontiasis is a disease prevalent among poor people in remote rural communities who obtain their drinking water from ponds and wells.
2. The intermediate hosts are various species of *Cyclops,* but only those which are predatory and live in stagnant water with a pH of 6.0 to 6.8 are likely to play a significant role in transmission. Hence people who obtain their water from flowing rivers or streams are unlikely to be infected.
3. This observation is confirmed by the distribution of the disease in southwestern Nigeria where the occurrence coincided closely with the source of water supply. Thus, for example, in Ilesha and Ile-Ife with numerous streams providing good drinking water, cases of the disease are extremely rate, while people in zones 1 to 7 who obtain their water from ponds are likely to have high rates of infection.
4. The transmission pattern depends on types of climate and amount of annual rainfall. In the Sahel savanna region of northern West Africa where the annual rainfall is less than 65 cm restricted to 3 or

4 months of the year, transmission occurs around the end of the rainy season when the pond water is available. For the rest of the year, the ponds are dry, water being obtained in most cases from bore-wells which are free of infection. In such cases, the incapacitating effect of the infection reaches its peak at the end of the following rainy season (since the pre-patent period is about 12 months). This coincides with the harvesting season.

In the more southern Guinea savanna areas with an annual rainfall of more than 150 cm, the ponds contain water all the year round. In the rainy season, however, there is very little transmission because the ponds are flooded and become turbid and unsuitable for cyclops. What is more, the people have an alternative source of supply of drinking water. In such areas, transmission is restricted to the tail end of the dry season when the ponds assume their role as

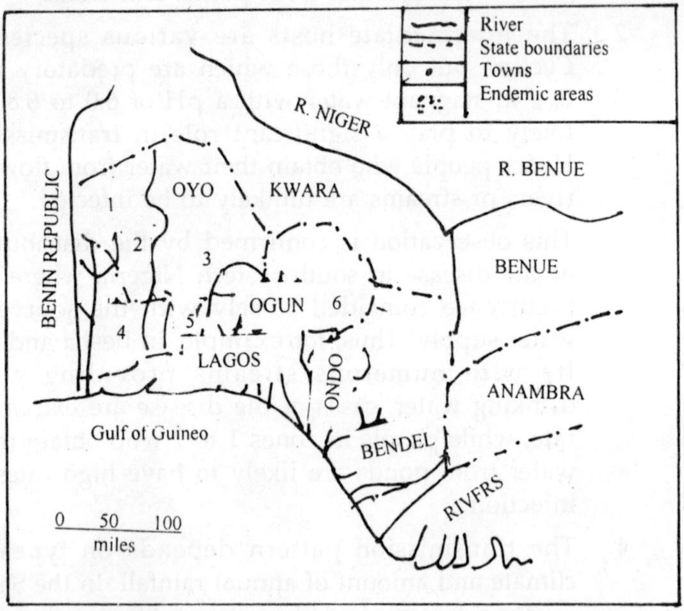

Fig. 5.5 : Distribution of guineaworm in south-western Nigeria.

transmission sites. Thus patent guineaworm disease reaches its peak in the following late dry season and early rainy season. This coincides with the main planting season. The seasonality of the disease has been studied by Onabamiro (1951, 1952c) and Lyons (1972).

5. The majority of infected people (over 60 per cent in Nigeria) have only one worm emerging at any given time, the mean being 1.17 to 1.69 in Ghana and 1.5 worms per person in Nigeria. However, cases of multiple infection are not infrequent, up to fourteen worms having been obtained from a woman in Nigeria.
6. It is not clear why higher figures are not obtained because people in endemic areas are highly exposed to infection. For example, Onabamiro showed that there are up to 152 cyclops in 5 litres of water in a village pond, of which about 5 per cent are infected with one or two guineaworm larvae. He estimated from this that a man in such a village would swallow between 75 and 200 infected cyclops per year through drinking pond water. With such a high level of exposure to infection, it is surprising that the intensity of infection is not much higher with more worms per infected person than observed. It is important to point out that Lyons (1972) found a much lower infective density of cyclops per litre in Wa, northern Ghana, a fact which may explain the lower incidence of the disease there.
7. The age and sex distribution of persons with infection varies a great deal. In the absence of national figures, reports from a few small-scale epidemiological studies will be given:

   (*a*) Overall incidence: 83 per cent in a village in south-western Nigeria; 53 per cent in four villages north of Ibadan 13.5 per cent in seventeen villages in Ibadan district; 26.1 per cent (range 1–39 per cent) in forty-three villages in Wa district of Ghana; and 1.2 per cent in 159 villages in Accra Plains, southern Ghana although they pointed out that this may be an underestimation since less than 1.0 per cent of infected people attend medical clinic.

(*b*) Age distribution: highest incidence among 15 to 40 year age group; 40 to 49, followed by 10 to 14 year age group, 20 to 24 year age group; and 25 to 44 year age group.

(*c*) Sex distribution: males at greater risk; equal rates of infection among males and females, although with higher incidence in girls than boys; significantly higher rates in females than males and the reverse.

8. A study of the water consumption pattern may explain differences in age and sex distribution in infection. Thus, heavy physical exertion in the farms in February to April in preparation for the planting season and in June to September (weeding and harvesting) which are thirst-inducing activities may result in the consumption of large quantities of water from ponds near the farms. This may account for the greater exposure and hence prevalence of the disease among the older adults (40 to 40) and the older children (10 to 14) who assist them in the farms, as observed by Kale (1977).

9. One remarkable feature of the disease is that in a community some individuals are susceptible to repeated infection every year while others, equally exposed, escape infection. Scott (1960) found 50 per cent of the population in an area of Ashanti Ghana infected by the age of 10 and 25 per cent never infected, and attributed this to the level of acidity in the stomach, those with high gastric acidity being more protected because the larvae are killed in the stomach, while those with low gastric acidity

being more prone to infection. Belcher *et. al.* (1975) tentatively offered this as one of the factors responsible for greater prevalences among male farmers because periods between meals are extended while working far away from home in the farms. This tends to reduce stomach acid and so enhances the chances of survival of larvae. However, Gilles and Ball (1964) found no significant relationship between infection and gastric acid secretion and in an experiment, found that the larvae were very sensitive to low concentrations of acid, 60 per cent of emerged larvae kept in 0.05 per cent HCI being dead within 1.5 to 3.5 hours.

10. There is no immunity to reinfection.
11. The period of disability (incapacitation or prostration) in untreated cases lasts for 5 weeks to 10 weeks. More importantly, it coincides with the planting or harvesting season in the community. In a situation in which over 50 per cent of the adult male farmers could be affected, it is easy to see what an impact this would have on the agricultural output and hence the economy. Also, the disease interferes significantly with school attendance by children, as shown in a study of a village in western Nigeria in which 18 per cent of the children were absent from school for up to 5 weeks.
12. There are no reliable figures about the annual loss to the national economy due to guineaworm disease; neither is it easy to estimate the cost among communities of self-employed peasant farmers. However, studies by Belcher *et. al.* (1975) in the Danfa Comprehensive Rural Health and Family Planning Project in southern Ghana give some indications as to the magnitude of the problem:
    (*a*) In southern Ghana, farming income is seasonal with 75 to 80 per cent of the total annual

income obtained from harvests in June to September, precisely the period when disability due to the disease reaches its peak.

(*b*) It is not easy during this period to secure alternative labour because:

(*i*) Other male members of the family or household who should be in a position to assist in the farms are also likely to be incapacitated at this crucial time.

(*ii*) Wives are engaged almost full-time on domestic chores—cooking, washing, fetching water, minding the children, trading, etc., and in any case can only do light work, *not* heavy farming duties.

(*iii*) Neighbours can not help either, because they have their own farms to look after, and/or are invariably afflicted as well.

(*iv*) Recruitment of casual migrant farm labourers is out of the question because thee is no cash to pay their wages.

(*c*) On the whole, the cost to the farmers is more than a month of labourers' wages.

(*d*) Quite apart from the direct loss, the prostrated farmers are incapable of increased agricultural production or of innovating improved farming technology.

GUINEAWORM—PREVENTION AND CONTROL

Dracontiasis is one of the most easily preventable parasitic diseases of man in tropical Africa for the following reasons:

1. The transmission foci are usually small and well defied.
2. The transmission period is limited to only a few weeks in the year.

3. There are no animal reservoirs.
4. Preventive measures, at least theoretically, involve no more than such simple procedures as:
   (*a*) Provision of a potable piped water supply or, alternatively,
   (*b*) Introduction of more hygienic methods of collecting and sorting water,
   (*c*) Purification of drinking water by simple boiling or filtration (even if only through a cloth) before consumption,
   (*d*) Prevention of contamination of water by active cases, and
   (*e*) Effective health education campaigns.

Unfortunately, however, provision of piped water to all the rural communities in endemic areas is not feasible because of the high costs involved, and results of health education campaigns in Nigeria have been largely ineffective. An interesting study by Ward *et. al.* (1979) on the impact of the attitudes, perception, and practices on management and prevention of the disease among members of five communities in the Danfa Project in Ghana revealed the following as the chief factors militating against the success of preventive measures:

1. The habit of farmers of drinking directly from ponds and filling their gourds en route to the farm complicates the whole process of prevention because it means that boiling and filtering of drinking water at home is no longer an adequate safeguard.
2. The belief that the disease is not preventable and that public discussion of the problem increases the chances of catching it are effective barriers to the success of health education campaigns.
3. Most patients do not attend modern health clinics, either because they are greatly incapacitated, often

immobile, or mostly only able to walk with a limp and in great pain, or because of reliance on traditional treatment. Furthermore, such treatment involves incision of the ulcer thereby increasing the chances of secondary bacterial infection and hence greater disability.

4. Above all, the general ignorance about the life cycle and mode of transmission, the lack of awareness and appreciation of the conceptual basis for the relationship between drinking contaminated water now and suffering from the disease one year later, and lack of awareness of the greater risk to adult males, demand great ingenuity and innovative ideas in devising persuasive and convincing health education strategies.

Chemical control of the vector is also feasible. Many insecticides have been found to be lethal to cyclops; indeed, 'were it not for its cummulative toxicity to other forms of life DDT would probably be the most satisfactory compound on a cost/effectiveness basis'. Care must be exercised in the choice of chemicals to be applied to pond water, because it may render the water unsuitable for consumption, not only by the people themselves but by their cattle as well. However, some organophosphorous compounds, in particular 'Abate' (=Temephos) have been shown both experimentally and in field trials to be effective and safe for controlling cyclops in the ponds. Although Temephos is being used effectively in the control of *Simulium* larvae in the onchocerciasis programme in the Volta River Basin, it is not yet in widescale use for guineaworm control.

In conclusion, most experts are agreed that the most effective preventive measure is the provision of a safe potable water supply. This has been well documented in the cases of Fiditi and Igbo—Ora in western Nigeria. Before the supply of piped water in 1963 and 1965 to these two towns (population 30000) the incidence of the disease was 20 per cent in school children in the former and 60 per cent in the general population in the latter. After, there was a dramatic fall in incidence, such that no further cases were reported after two years even though the disease was still endemic in the surrounding areas. If a low-cost system of water supply could be developed, not only would guineaworm be eradicated but the incidence of other water-borne diseases like gastro-enteritis, typhoid fever, hepatitis, poliomyelitis, and cholera, etc., in rural tropical Africa would be greatly reduced within a year or two.

Distribution and Importance

Infections in rats are widespread in tropical areas of the Eastern Hemisphere such as the Malay Peninsula, Japan, the islands of both Melanesia and Micronesia, northern Australia, Hawaii, and East Africa including Madagascar. According to Alicata and Jindrak (1970), the original enzootic area of *A. cantonensis* was Madagascar, the large is land off the southeastern coast of Africa, from which it spread northward to India and Sri Lanka, thence to the Malay Peninsula. Migration then took place in a number of directions: northward to China, eastward through the islands of the South Pacific, and southward to New Guinea and Australia.

Infections in humans have been found in Southeast Asia, Japan, many islands of the South Pacific, and Hawaii. All well-documented infections have been found to be contiguous with the distribution of the infection in rats.

Although infections are widespread in rats, infections in humans are sporadic in the present enzootic area and they are related to types of food eaten and their method of preparation. Infections in humans seldom result in death; it is therefore difficult to determine with certainly whether *A. cantonensis* was actually the cause of eosinophilic meningitis in any specific instance. Most often, only a few persons suffer from the infection at a particular location and time.

Morphology

The female worms are 18 to 33 by 0.28 to 0.5 mm long with the vulva about 0.25 mm from the posterior end. Males are 15 to 22 by 0.25 to 0.35 mm and have subequal spicules that are 490 to 1200 μm long. Both sexes feed on blood and therefore have a red intestine around which the reproductive system winds; this gives a barber-pole appearance also noted in such worms as *Haemonchus*.

Life Cycle

The life-cycle described here is that which occurs in the normal hots, rats and in the invertebrate hosts. As is true with many other metastrongyles, the L_3s are ensheathed in the cast cuticles of the two previous stages. The L_3s are eaten by rats and within a day reach the central nervous system through the circulatory system. The early adult stage is reached in 11 to 13 days after infection, and migration

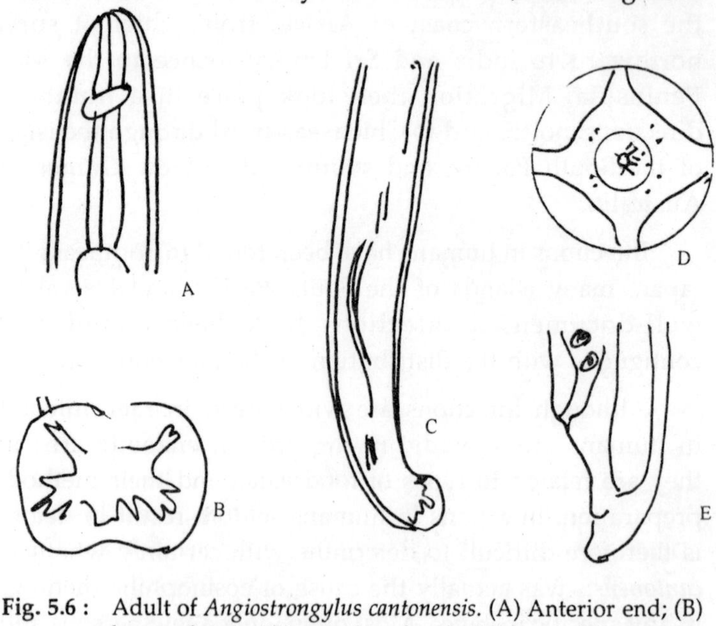

Fig. 5.6 : Adult of *Angiostrongylus cantonensis*. (A) Anterior end; (B) bursa of male; (C) spicules of male; (D) head: (E) posterior of female.

then begins. The worms move to the meninges covering the brain and remain between the layers (subarachnoid space) for about 2 weeks. They then move to the venous system and are carried through the heart to the pulmonary arteries, where they remain and begin laying eggs. The unembryonated eggs are carried with the arterial blood to the capillaries in the alveoli of the lung, where they lodge. The L_1s break out of the alveoli and pass up the respiratory tree and are swallowed; the pre-patent period is 42 to 45 days, with larvae appearing in the faces at that time.

The first intermediate host is a snail or slug, usually a terrestrial one. The host range is quite broad, but development takes place more readily in some species than in others. The L_1s penetrate the foot of the mollusk and reach the L_3 in about 18 days at 21 to 25°C.

Having reached the infective stage, the I_3 organism can reach the definitive host by one of the following means:

1. The gastropod can be eaten.
2. The L_3s can escape from the gastropod and be eaten with other food or with water.
3. A transport host can be infected by either of these routes and then it, in turn, may be eaten by the definitive host. The most important transport hosts seem to be freshwater crustacea such as prawns and crab, and land planaria. The most frequent mode of infection in wild rats is not known.

The life cycle in humans seems to be the same as in the rat except for the final site. In humans the worms remain in the brain, lyiııg either between the meninges or migrating through the brain tissue itself.

Diagnosis

Diagnosis is difficult in humans; in the other hosts, an examination may be done to find the worms in the pulmonary arteries. Unless the person has died, the diagnosis of *A. cantonensis* infection is indirect and is based on the following:

1. Clinical signs and symptoms
2. Food habits and a history of having eaten a likely source of larvae about 2 weeks previously
3. The presence of eosinophils in the cerebrospinal fluid (CSF)
4. Evidence of antibody production.
5. Finding worms in the CSF (rare; only two instances are on record).

There are many causes of meningitis; therefore both a skilled diagnostician and a laboratory capable of isolating

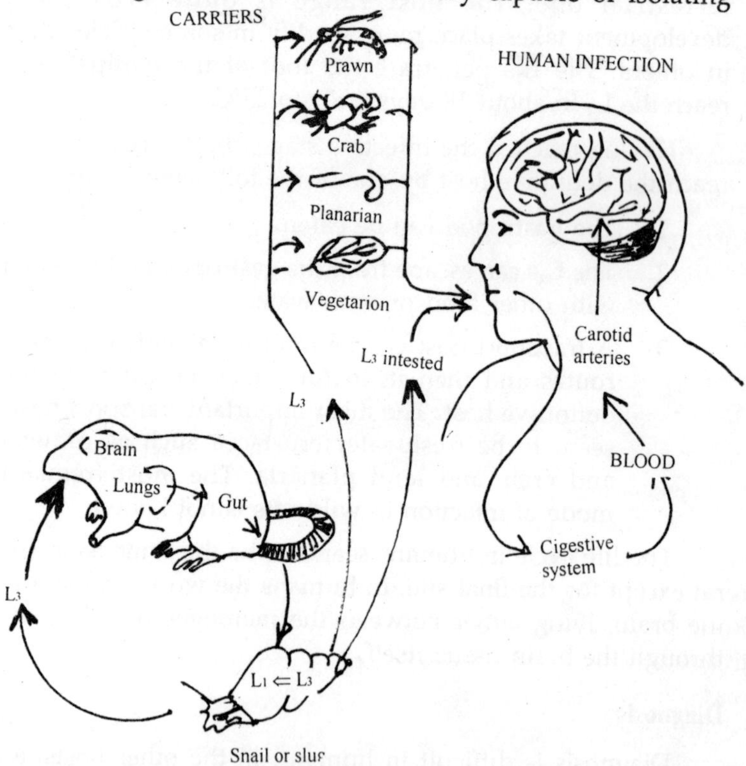

Fig. 5.7 : Diagram of the life cycle of the rat lungworm *Angiostrongylus cantonensis,* and avenues of human infection.

and identifying various infectious agents are required to determine whether bacterial, viral, or other parasitic agents may be the cause of the disease. Skin and serological testing can indicate whether a nematode may be involved in producing antibody. A further narrowing of the possible causes of meningitis comes from questioning of the patient. In an enzootic area, the typical symptoms coupled with suggestive laboratory tests and food habits consistent with becoming infected provide a good indication of infection with *A. cantonensis.*

Other infections with which *A. cantonensis* can be confused are *Taenia solium* cvsticerci *Paragonimus westermani,* and *Gnathostoma spinigerum*. Noninfectious processes such as cancer and multiple sclerosis may also cause an increase in eosinophils.

Treatment

Thiabendazole has been used with some success in treating laboratory animals. This drug has been suggested as appropriate for treatment of some *A. cantonensis* infections in humans. Since thiabendazole has a low degree of toxicity, harming the patient by the drug is unlikely.

Some reservations have been expressed on the desirability of treatment in some instances and on the level of drug that might be administered to a patient. The concern is that a worm migrating through the brain tissue clearly causes damage, but a dead worm becomes a foreign body against which a strong leukocytic reaction is directed. The result is a large granuloma, which itself causes damage. Roughly the same hesitation is expressed on the protocol of treatment; if a high level of drug is used, all of the worms may be killed at once, producing a toxic reaction that could be detrimental to the patient. Although some cases are on record in which persons have died from the infection, most recover completely without specific therapy. The physician's cardinal principle pertains first, do no harm.

Table 5.1. Clinical Signs and Symptoms of Eosinophilic Meningitis and Presumably Infection with *A. cantonensis*.

Sign or symptom	*Percentage having*
Headache	90
Stiff neck	56
Vomiting	56
Paresthesia (abnormal sensation)	54
Fever	41
Paralysis or tremors	9

[Modified from Kuberski and Wallace, 1979].

Host-parasite Interactions

Frequently, but not invariably, *A. cantonensis* has a low degree of pathogenesis in the rat. On the other hand, in humans there is a full range of effects including severe damage to the central nervous system leading to death.

In humans it is the consensus that there are many inapparent infections and others in which there may be only mild symptoms of meningitis but with eosinophils in the cerebrospinal fluid. Lesions are related to both the location of the worms and the number in the nervous tissue. Eosinophilia indicates an allergic response to the worms and the meningitis indicates that there is interference with normal nerve impulse propagation and probably swelling of the CNS. If damage is sufficient, paralysis of particular parts of the body or of areas served by the cranial nerves may result.

Epidemiology and Control

The life cycle is probably maintained in rats by their eating infected snails, but freshwater crustacea are a second source.

The food habits of people in various localities or their ethnic derivation determine whether infection is likely to occur. This is another example of a parasite that cycles silently in the non-human population and can be transmitted when humans come into contact with the cycle. In this case, food preferences are the key.

Among certain peoples of the Pacific islands, foods such as tiaro on Tahiti or keloguen on Guam consist principally of grated coconut mixed with the juice from ground or crushed freshwater prawns. On some of the Pacific islands, 5 to 10 per cent of freshwater prawns have been found to be infected, mostly with two or three larvae of *A. cantonensis* but in one instance with over 3000.

The predilection of Europeans or North Americans for green salads leads to infection. Terrestrial gastropods, and planaria secondarily, shed infective larvae onto plants in the field and the larvae are then eaten with the salad. In some enzootic areas, eosinophilic meningitis occurs in Caucasians and not in the natives because the latter do not usually eat leafy salads.

In a case report a man suffered typical *A. cantonensis* signs and symptoms following ingestion of live slugs. The 59-year-old farmer had back pain, and Chinese physician had recommended eating raw slugs, which he had done for about 4 years, eating from five to ten daily. It is not stated whether the diet of slugs alleviated his back pain.

Since eosinophilic meningitis is usually a sporadic infection in small numbers of persons at any one time, control programmes have not been developed. At the public health level, there are two approaches that might be considered:

1. Rat control
2. Education to alter food habits.

Since the definitive host range is relatively narrow, rat control might be worthwhile, but in general rats are not restricted to areas of human habititation. Control is therefore likely to be difficult.

The other alternative, altering food habits, may be equally difficult. When it is explained to people what might happen if they eat tiaro or a green salad, the response is likely to be, "yes, what you say is probably true"; then they will continue to eat what they always have.

Drinking untreated surface water is also a hazard since larvae escape from the intermediate host or from planaria and may be transmitted if this happens in small streams.

At the public health level, probably the best that can be done is to monitor the extend of the infection in the intermediate and transport hosts and make available information on the consequences and likelihood of infection. The responsibility then falls on the individual on whether to avoid infection or not.

It is also important to note that the *larvae* of several ascaridoid species-notably *Toxocara canis* of cats and dogs can survive in abnormal hosts (such as man) wandering through tissues as a *larva migrans* which often causes serious pathological effects.

GENUS ASCARIS

Type example: Ascaris lumbricoides

The best-known species of this genus is the so-called 'large roundworm', *Ascaris lumbricoides,* which is a parasite of man, some apes and pigs all over the world. Much debate has taken place concerning the taxonomic status of the human and pig *Ascaris*. Some authors have claimed to detect small morphological differences between worms from these different hosts; others have failed to confirm these differences. Scanning electron microscopy has, however, confirmed that differences in the shape of lips and denticles do occur. It is probable that they represent physiological 'strains' of the same species (cf: *Echinococcus; Entamoeba; Trichinella*). Some workers, however, refer to them as separate species, *A. lumbricoides* (man) and *A. suum* (pig), or subspecies *A. lumbricoides lumbricoides* and *A. lumbricoides, suum;* the first mentioned nomenclature is followed here. Antigenic analysis has shown that *A. suum* shares eleven antigens with *A. lumbricoides,* but in addition, has three specific antigens. Although it is often said that the pig 'strain' will not develop in man and *vice-versa,* there is now unequivocal evidence from experimental and accidental laboratory infections that the 'pig' form can indeed infect man and the 'human' form can develop in pig. The infection rate of the heterologous species, in each case, appears to be low and there are also differences in the rate of development. Nevertheless, it is important to take note that it is possible for the eggs of the pig *Ascaris* to infect man, for this material is often used for experimental or dissection purposes in parasitological laboratories. In addition to pigs and man, *A. lumbricoides* has been reported from other hosts such as bovines and squirrels.

External Features

Females (200-350 mm) are much larger than the males (150-310 mm) and the latter are also distinguished by a curved posterior end bearing a slit-like anal opening from which often protrudes a pair of copulatory spicules. The cuticle, which bears transverse markings, is sometimes brown in colour due to the presence of quinone-tanned proteins. The

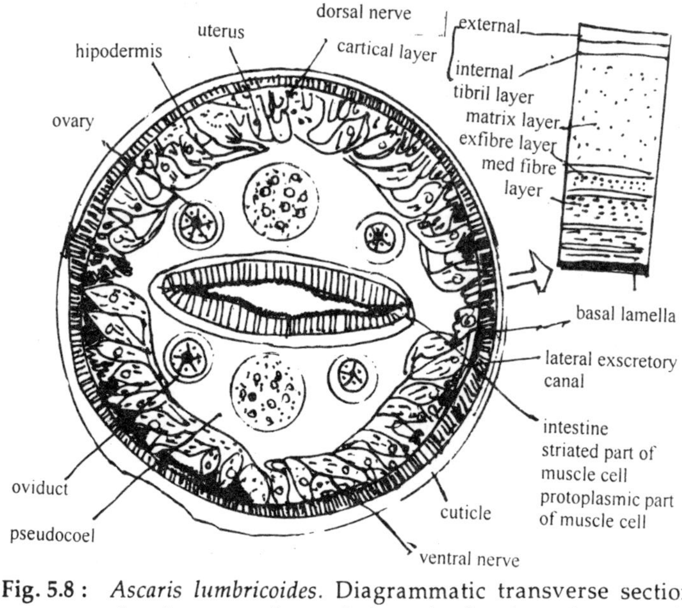

Fig. 5.8 : *Ascaris lumbricoides.* Diagrammatic transverse section showing general morphology the figure on the top right is an enlarged section of cuticle.

position of the excretory canals is marked by two broad brownish lateral lines, and the position of the main nerve cords is similarly marked by dorsal and ventral lines just visible as white lines through the cuticle. Some 2 mm from the anterior end is an excretory pore, ventral in position and visible only in exceptional specimens. In the female, a narrow constricted band may be seen about one-third the distance from the anterior end, and this bears the slit-like opening of the vulva. In the male, there is a sub-terminal cloaca.

Digestive System

The mouth is surrounded by three cuticulate lips which are best seen in *en face* sections. The dorsal lip bears a pair of papillae which represent the two papillae of the external circle. The paired subventral lips bear (*a*) amphids, and (*b*) fused ventro- and latero-ventral papillae. The inner edges of the lips are provided with dentigerous ridges whose structure is believed to differ in the human and pig varieties. The alimentary canal is the usual nematode type. As in other species ultra-structure studies have shown the presence of well developed microvilli. In the horse species (*Parascaris equorum*) the presence of the latter has been calculated to increase the surface area for absorption by 129 times.

Sections of the gut have revealed that the food of *Ascaris suum* contains a high proportion of cells which have been sloughed off from the intestinal mucosa of the pig.

Muscular System

This is built on the polymyarian plan with about 150 muscle cells in each quadrant. The vacuolated protoplasmic parts of the muscles are particularly well seen in transverse sections.

Pseudocoele

This cavity containing a protein-rich fluid of characteristic unpleasant odour, which sometimes induces allergic symptoms in those frequently handling specimens.

Reproductive System

The female system is double with thread-like ovaries passing into oviducts and finally into wider uteri opening at the vulva. The male system is single with a large seminal vesicle joining the rectum and spicule pouch just before the cloaca. The caudal end of the male bears some fifty or more pairs of simple pre-anal papillae and four pairs of single post-anal papillae, one pair of double post-anal papillae and one pair of phasmids.

Excretory System

This is often described as being of the ∩ type, but more recent studies suggest that it is of the H type, that is, with distinct anterior canals. There is probably a nucleus present in the short terminal duct leading to the excretory pore and another, *the giant sinus nucleus,* associated with the left lateral canal. Four coelomocytes occur in the body cavity situated anterior to the vulva, as shown in the Fig. 5.9.

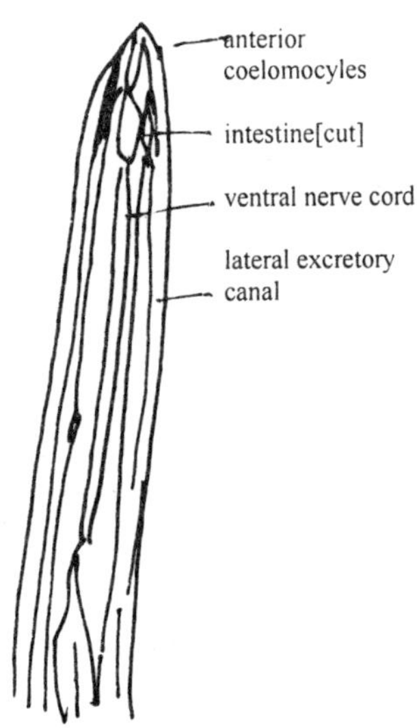

Fig. 5.9 : *Ascaris lumbricoides*—anterior end dissected.

Life Cycle

When eggs are passed in faeces, their further development is largely dependent on the oxygen tension, moisture content and temperature of their environment. In a cold dry atmosphere, it is said that eggs may remain viable for up to six years and in infected areas they may be widely distributed. At a temperature of 22-33°C, the eggs give rise to coiled rhabditoid larvae within 9-13 days, but do not hatch (except accidentally) until taken into the definitive host. The shell layers of the egg provide a remarkably resistant structure which can withstand many chemical agents, and in the laboratory eggs are normally embryonated experimentally in 2 per cent formal solution.

For accounts of the life cycle of the pig *Ascaris* see Douvres, Tromba, and Malakatis (1969) and Galvin (1968).

Pathology of Ascaris Infection

The pathological effects of ascariasis can be distinguished into two groups: the effects due to the adults and those due to the migrating larvae.

Effects of Adults

Light to moderate infections present minor or no symptoms. It is believed that the consumption of semi-digested food from the host may contribute to malnutrition and underdevelopment in children. To what extent, however, has not been fully measured, especially in circumstances of multiple infection with other intestinal parasites and among people in which under and malnutrition are already rampant. However, allergic responses to toxic waste products of the parasites are evident in most patients. These present in the form of abdominal pain, rashes, insomnia, restlessness, etc.

Cowper (1966) has briefly reviewed the effect of *Ascaris* infection on metabolism and nutrition. It appears that adult worms interfere with normal metabolism through:

1. Blockage of lumen, reducing the absorptive area.

2. Absorption of the intestinal contents.
3. Ingestion of blood by destruction of intestinal mucosa.
4. Probably by neutralizing pepsin and typsin with an anti-enzyme, ascarase.

He also reports that ascariasis interferes with protein digestion in children and that, in combination with hookworm infection, 'cold cause, accompany or accentuate kwashiorkor in African children'.

Heavy infections produce similar but more marked effects. In some cases, due to circumstances not yet understood, the worms knot together to form a tangled mass completely blocking the intestinal lumen with sometimes fatal consequences. Also, worms can penetrate through the intestinal wall or appendix into the body cavity, causing peritonitis which can be fatal. Cowper (1966) cites the case in which *Ascaris* was incriminated as the cause of death in 30 per cent of forty-one post-mortems in Lagos, Nigeria. He also mentions cases of heavy infections, many patients as Ibadan, Nigeria, harbouring over a hundred worms and one patient as many as a thousand.

Adult worms produce pathological effects when they wander into abnormal sites. Thus adults worms have been

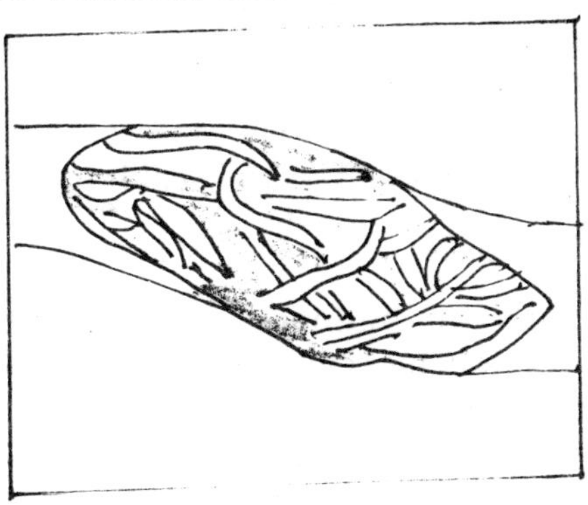

Fig. 5.10 : *Ascaris* blocking intestinal lumen.

found clogging the appendix, bile and pancreatic ducts, trachea, eustachian tubes, and urino-genital ducts with grave consequences. A case of urino-genital ascariasis was described in the urethra of a 2½ year old boy in Kenya by Pamba and Musangi (1978). Worms can also produce bizarre effects when they migrate upwards to the stomach, the resulting nausea causing the host to vomit live worms through the mouth. They can also emerge involuntarily through the nose, mouth, or anus to the chagrin and embarrassment of the unsuspecting host.

Effects Caused by Migrating Larvae

When the larvae break out of the lung capillaries, they cause haemorrhage, and the accompanying accumulation of polls of blood, dead tissue cells, etc., lead to congestion causing focal pneumonitis. Subsequent inflammatory reaction produces *Ascaris* pneumonia. Also several of the lesions caused by these larvae predispose the lungs to secondary bacterial infections which could be fatal. Indeed, the frequency of pulmonary tuberculosis in West Africa has been attributed to this. Finally, some migratory larvae get lost and die in abnormal locations like the brain, spleen, liver, etc., where they induce inflammatory responses which could be confused with other conditions.

Thus, far from being a harmless commensal, a belief which is widely held in tropical Africa by communities where the high prevalence leads people to take *Ascaris* infection for granted, ascariasis is a serious cause of morbidity and mortality and is an important public health problem.

Diagnosis and Treatment

It is difficult to diagnose ascariasis by clinical manifestations or by the identification of the migratory larvae in early infection. Definitive diagnosis is by identifying adult worms passed out spontaneously by the host or by demonstration of the characteristic eggs in faecal samples. Eosinophilia of 30 to 40 per cent can be detected, but this is a common feature or helminth infections.

Several drugs effective against *Ascaris* are available. The drugs of choice are piperazine salts and pyrantel pamoate, single doses of which can procure 80 to 90 per cent cure rates. These drugs are safe and easy to take. Kale (1977) tried a mixture of pyrantel pamoate and oxantel pamoate on Ibadan children with some success. Other drugs are levamisole and mebendazole. Lechat *et al.* (1974) obtained a significant reduction in prevalence and incidence of infection with *Ascaris* and hookworms with indiscriminate mass treatment with levamisole in rural areas of Zaire. Drugs like tetrachloroethylene, used for treatment of hookworm infection, cause *Ascaris* to knot up with resultant grave effects. In cases of mixed *Ascaris* and hookworm infection, it is standard procedure to expel *Ascaris* first before administering the hookworm expeller. There does not seem to by any effective treatment for migrating larvae.

Prevention and Control

This depends on perturbation of some parameters of the faecal-oral transmission cycle. An account of epidemiology and control of ascariasis has been given by WHO (1967). A detailed discussion of epidemiology and control of mixed nematode infections in tropical Africa is given below.

Epidemiology

Infection being acquired through the direct ingestion of eggs, epidemiology is intimately linked with the level of sanitation in the community. The lack of latrines and adequate sewage disposal facilities result in the spread of eggs, bringing about widespread contamination of food and drink. Thus people become infected by eating food (including raw leafy vegetables) contaminated with eggs and drinking contaminated water. Infection can also be acquired through contaminated, unwashed fingers. Eggs can be spread by flies and other animals, circulation of bank notes, and by wind during dry conditions.

One feature which has a profound influence on the epidemiology is the resistant capacity of the egg which can embryonate under the most adverse environmental conditions and survive in the presence of some of the chemicals known to be lethal to other parasite eggs. This capacity is conferred by the ascaroside layer of the egg shell and makes it possible for the eggs to remain viable in the soil for extended periods (up to ten years has been reported).

Country	Locality	Size and type of sample	Infection rate						Reference
			Ascaris	Hookworm	Trichuris	Strongylordes	Enterobius	Laama	
1. Senegal	Serere-speaking community	257 faecal samples from urban area							Juminer et a, 1971
2. Ovary Coast	d' Adzope region	68 children under 14 years							Nozais et a, 1975
3. Upper volta	North-West region	7,374 stool samples from hospitals							Gendror, 1975
4 Nigeria	1 Southern Negeria	110,431 stool samples from 9 urban							Hinz, 1966
	2 Nsukka E. Nigeria	529 students and kitchen staff, Univ.							Okpala, 171
	3 Benin-City Bendal State	6,213 stool samples at a hospital							Obramiwe, 1977
	4 2 Yoruba villages W.Nigeria	90% of children under 15 years: under 15 yrs. over 15 years							Gurney, 1969
	5 Legos S Nigeria	515 stool samples of Govt. workers							Okpala, 1961
	6 Lagos State	5,595 primary school children							Ejezie, 1981
	7 Ibadan W Nigeria	i) 380 school children(6-15 years)							Onakamiro, 1957
		ii) 140 children from elite school							
		iii) 1,125 children from lower social class							Oduntan, 1974
		iv) 12,384 hospital patients in 1970							Ogunba, 1974
	8 U.C.H. Ibadan	21,700 routine stool samples							
	9 Akuffo village, near Ibadan	500 villagers							
	10 Lagos	4,700 children and adolescents							Cowper, 1966 and 1967
	11 Lagos	500 stool samples							
	12 Jos, Plateau State	7,000 stool samples							
	13 Katsina Kaduna State	i) 536 Habe ii) 236 Fulanis							

Fig. 5.11 : Prevalence of Human Intestinal Nematodes in some Tropical African Countries.

In tropical Africa, many surveys have revealed high infection rates of over 50 per cent. Furthermore, *Ascaris* infection is found to be associated with *Trichuris trichiura* and hookworm infections.

TRICHURIS TRICHIURA

Trichuris trichiura infect the lower part of the alimentary tract—the caecum, lower ileum, and appendix of man—particularly in tropical countries. The common name, whipworms, describes the shape of the body which is like a whip—a log, thread-like portion with a thick handle. But the generic name *Trichuris* is a misnomer because it implies that the thread-like parts is the tail.

ALIMENTARY CANAL

Buccal Cavity

This is variable in shape and size, and in degree of differentiation. In some forms, its cuticular lining may be moulded into rods or plates, often bearing teeth. The latter are particularly well developed in intestinal forms such as hookworms which rely on sucking blood from the mucous membrane.

Pharynx (Oesophagus)

The pharynx is one of the most characteristic features of nematode morphology. Although the terms 'pharynx' and 'oesophagus' are used as being synonymous, *pharynx* is used in preference to *oesophagus* as it is used for homologous structures in related groups. Since nematodes generally appear to have a high internal pressure, and since the intestinal wall of nematodes is only one cell chick (and therefore, liable to collapse under turgor pressure) the intestine can only be filled by a powerful pumping organ. The pharynx which serves as this organ has a syncytial structure with a triradiate lumen. It is lined with cuticle, continuous with that of the buccal cavity and is, therefore, part of the stomodaeum; the cuticular lining is shed at ecdysis. The median and posterior regions of the pharynx are generally swollen into one or two muscular bulbs, but these are lacking in some nematodes. The pharynx shows considerable variation, both in its structure and mode of operation, a diversity correlated largely with feeding habits, chief of which are microbivorous (*i.e.* on bacteria), saprophagous (*i.e.*, on decaying organic material), phytophagous (*i.e.* on fungi, algae and cells of higher plants) or carnivorous (*e.g.* on Protozoa). Although most of these terms refer to free-living nematodes, it must be pointed out that many nematode parasites have free-living stages.

Most nematode parasites of vertebrates feed on the semi-liquid contents of the alimentary canal, the tissues of the mucosa, lysed tissues of other organs, or blood or other body fluid and their mouth structure and pharynx are appropriately adapted. For detailed consideration of feeding and digestion see Lee (1965*a*) or Crofton (1966).

In addition to muscular bulbs, many nematodes possess a well-developed system of valves which prevent cuticularized. Caudal alae or cuticular bursae commonly present. Male with paired genital papillae. Six (or less) coelomocytes. Caudal glands absent. Saprophages in soil (rarely water) and parasitic in vertebrates and invertebrates.

Rhabditida

Small transparent meromyarian worms. Pharynx usually with a posterior bulb and frequently a prebulbar swelling. Mouth with three to six lips. Females oviparous or viviparous and may be parthenogenic or hermaphroditic. Majority free-living, some with both free-living and parasitic phases (*e.g. Rhabditis, Strongyloides*).

Ascarida

Mouth with three lips (one dorsal and two sub-ventral); pharynx bulbed or cylindrical; vagina elongate; male usually with ventrally curled tail and two spicules; alae may be present (*e.g. Ascaris Enterobius*).

Strongylida

Male with copulatory bursa; mouth simple without lips, frequently with a buccal capsule and teeth; usually meromyarin; pharynx muscular, club-shaped or cylindrical (*e.g. Ancylostoma, Nippostrongylus, Syngamus*).

Spirurida

Pharynx composed of two parts, both cylindroid, an anterior muscular portion and a posterior grandular portion; males usually with two spicules and well-developed alae on spirally coiled tail; mouth with no lips, rudimentary lips, or with two or four paired lips. Long, thin nematodes, parasitic in vertebrates in adult stage. Mainly viviparous (*e.g. Wuchereria, Dracunculus*).

TYPE EXAMPLE: RHABDITIS MAUPASI

A number of the most readily obtainable nematodes belong to the genus *Rhabditis*. They occur in soil, water and decaying organic materials, and are faculative parasites in the larval stage. *R. maupasi* occurs commonly in earthworms. Other common species in the same host are *R. terrestris, R. pellio* and *R. anomola,* but it is difficult (even for those working on this group) to identify the species from the female or protandrous hermaphrodite. Males may be rare in some species. Young encysted forms (sometimes known as *'dauer'* larvae, although the term is used rather differently by various authors) occur in the ovoid brown bodies lying in the coelomic spaces of the posterior somites. Free forms may usually be found in the seminal vesicles or coelomic spaces and are especially common in the nephridia.

In the earthworm, rhabditids never develop beyond the juvenile stage, presumably due to the lack of certain growth factors, but with the available nutriment, are just able to survive without further differentiation. When an earthworm dies, however, the decaying flesh, and the bacteria which thrive on it, provide additional food to raise the nutritional level to a degree which permits the juveniles to reach sexual maturity within a few days. Flourishing laboratory cultures are readily obtained by allowing a portion of the body wall of an earthworm containing nephridia, to undergo degeneration on nutrient agar plates.

External Features

Both sexes are cylindrical and elongated with a very thin almost hair-like caudal termination or post-anal tail. The males (which in some 'races' are rare in laboratory cultures) are characterized by possessing a flared posterior end forming an accessory copulatory organ.

Body Wall

A thin transparent cuticle covers the body. This is clearly secreted by the hypodermis, a subjacent layer containing a small number of nuclei. Beneath the hypodermis is a muscular layer consisting of a single sheet of longitudinal cells. *Rhabditis* is a *meromyarian* nematode, that is, it has few muscle cells in each sector. These muscle cells are unusual in that they have processes extending to the motor nerves instead of nerve processes extending to the muscles.

Digestive System

The intestine has a fine basement membrane supporting a single layer of cells. The pharynx and rectum have a cuticle, continuous with that on the exterior, which is shed at ecdysis. The mouth is enclosed in one dorsal and two ventral lips, each of which is further subdivided into two lobes, bearing a pair of sensory papillae. The narrow buccal cavity is heavily cuticularized and bears a ridge-like construction near its posterior end; the ridge bears fine cuticular teeth of taxonomic value. There is a muscular *pharynx* (=oesophagus) with a wide anterior region or *corpus* leading to a middle *pseudobulb,* connected by a narrow *isthmus* to a posterior *end bulb,* containing a tripartite valve; the bulb wall contains three gland cell nuclei. The arrangement of these cells, which ramify throughout the eosophagus, is of taxonomic importance.

Excretory System

The so-called 'excretory system' consists of a pair of lateral canals extending from the pharyngeal pro-corpus region nearly to the end of the worm, where they end blindly. These canals lead into an excretory sinus, which opens by a cuticular duct to the excretory pore. Also connected with the excretory sinus is a pair of sub-ventral cells. The excretory sinus bears a single nucleus; it is usually considered that the lateral canals are extensions of the sub-ventral cells.

Nervous System

The general features of the nervous system are very similar in all nematodes. Only those parts easily seen in living unstained specimens will be described here. A nerve ring, composed chiefly of nerve fibres, surrounds the isthmus portion of the pharynx. Nerve cells lie in contact with the nerve ring and dorsal and ventral cords join it.

Male Reproductive System

There is a single tubular testis reflexed at is anterior end and filled with spermatozoa in all stages of development. Spermatozoa are spherical and tail-less, a typical nematode feature. A well-developed vas deferens leads from the testis to the cloaca, in association with which are cuticular spicules

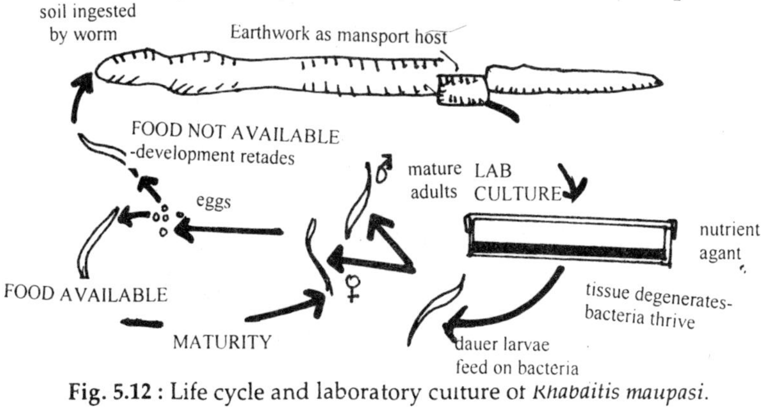

Fig. 5.12 : Life cycle and laboratory culture of *Rhabditis maupasi*.

with characteristic knobbed ends. These spicules appear to act as holdfast organs during copulation and are moved by protractor and retractor muscles. The tail of the male is further differentiated by being drawn out into longitudinal ridges—the *caudal alae.* These are thin-walled extensions of the postero-lateral body margins and also assist in holding the female in apposition during copulation. They are supported by nine pairs of *genital papillae,* finger-like projections of the ventral surface. Caudal alae are sometimes referred to as 'bursae', but this term, strictly speaking, is reserved for the wide caudal alae of the Strongylida.

Female Reproductive System

This is built on the typical nematode plan with paired ovaries converging on the vulva. The ovaries contain developing oocytes packed tightly together. The ends are retro-flexed and each becomes constricted to form an oviduct in which spermatozoa may be found. Fertilization takes place in this region, and eggs accumulate in the paired uteri, which connect by a short ventrally directed vagina, to the vulva.

In 'races' which are protandrous hermaphrodites, the worm has a genital system of the female form. Spermatozoa are first produced, then ova, which are fertilized by the stored spermatozoa.

Life Cycle

This is shown diagramatically in Fig. 5.12 When an earthworm dies, the contained third-stage 'dauer' larvae feed on the bacteria and fluid of the decaying organic material and develop into adults. These reproduce either bisexually or hermaphroditically and larvae produced from hatched eggs migrate into the soil to penetrate another earthworm via the nephridiopores or other apertures. Although the coelomic forms encyst, those in the nephridia and other sites remain free but, presumably due to the absence of suitable nutrients, they do not grow beyond the 'dauer' larva stage.

Chapter—6

Trichinella Pseudospiralis

Although raptors are exposed to a greater number of parasitic infections, little is known about whether such parasites affect the competitive fitness of these birds. Natural infections with the nematode *Trichinella pseudospiralis* are known to cause severe debilitation and death in Cooper's hawks (Wheeldon *et al.* 1983). Laboratory studies suggest that essentially subclinical infections with this parasite can elicit profound behavioural changes in American kestrels (*Falco sparverius*). The major effects of infection in non-breeding female kestrels involve a reduction in mobility. Severe and long-lasting effects appear as early as 15 days post-inoculation and coincide with the release of larvae into the tissue of the host. *Trichinella pseudospiralis* infections also severely impair the reproductive success of captive American kestrels. This is manifested in a high proportion of broken eggs, increased embryo mortality, and nest abandonment. The present study explores the behavioural dimension of this failure and examines the behaviour of infected nesting as well as nestling–parent interaction. Conceivably, behavioural patterns may be as sensitive to the effects of parasitic infections as are some secondary sexual characteristics. Finally, the study raises the question whether contact with infected individuals may alter the behaviour of other members of the family group.

Methods

American kestrels were bred and maintained in captivity according to the procedures of the Macdonald Raptor Research Centre of McGill University as described by Bird (1985). Birds were housed in cages measuring 2.4 × 1.2 m × 2.4 m (*L* × *W* × *H*). The cages consisted of a plywood frame covered, top and bottom, with wire mesh, and were equipped with shelters, rope and wood perches, as well as a plywood nestbox. One-way glass windows gave a view of the interior of the pen and the next box. A door provided access to the pen from a darkened, central observation corridor. Males were allowed to establish themselves in the cage for several days before being randomly paired with females. In the pens, kestrels were exposed to natural daylight and temperatures. Birds were fed freshly-thawed, 1–day–old chick enriched with SA–37 vitamin-mineral supplement.

At the time of pairing (mid-April), 13 randomly selected pairs of birds were inoculated by gavage with 100 larvae of *Trichinella pseudospiralis* per gram of body weight. A control group of 26 pairs was sham inoculated with 0.85 per cent saline. Half of the clutches of control parents subsequently received 100 larvae per gram body weight at the age of 2–5 days.

At the end of the experiment, infected adult birds were euthanized with CO_2, and individual total muscle burdens of larvae were obtained by artificial digestion of the skinned and eviscerated carcasses. Tissues were digested in 0.5 per cent pepsin for 2 h, filtered through 3 layers of cheese-cloth, and washed in 0.85 per cent saline.

Activity Budget

Information was gathered on the manner the behaviour of the kestrels was partitioned among various activities in time, as well on the relationship of these activities to the various stages of infection. An activity budget was obtained by recording the specific activities of individual kestrels during 15-min sessions at 15-s intervals (total of 60 observations per session). For convenience, the mean frequencies of each behaviour (out of 60) were converted to percentages. The method used was modified after Wiens *et al.* (1970) and Altmann (1974). All kestrels within the same flight cage (one family group) were observed simultaneously. Individual birds were identified with coloured plastic bands on both legs, as well as a single numbered aluminium band. At the age of 2–5 days, young nestlings were marked individually with ink on the down of different body regions. The behavioural data were processed through PROC RANK (SAS 1982) to obtain the ranks of the response variables, and parametric procedures were applied to these ranks. Since sample sizes were unbalanced, the General Linear Model (GLM) procedure (SAS 1982) was used instead of ANOVA. The GLM model examined the main effect of class variables (treatment, sex, time of the day) as well as crossed effects (treatment and time of the day, sex and treatment and time of the day, sex and time of the day). Duncan's Multiple Range test (DMR) was used to determine significant differences between the means of the ranked behavioural data. When interactions between the factors included the treatment, and were statistically significant, these are presented in the results. Thus, interactions between sex and treatment are designated as (S*T), between treatment and time of day (T*TD) or both (S*T*TD). Levels of activity are presented as means (X) of unranked data. Standard errors (SE) provide an indication of the variability about these means. The above two statistics were not used to determine significant differences between means, but rather provided a convenient format for quantifying the levels at which various behaviours occurred. Sample sizes (*N*) represent the number of 15-min bird sessions of observation.

The behavioural activities of adult kestrels were recorded throughout the periods of courtship, incubation and parental care. Their offspring were observed throughout their nestling and juvenile phases. The courtship period was considered to the interval between pairing and the time at which three eggs had accumulated in the nest. For kestrels, the task of incubation does not begin until the clutch is complete, or nearly complete. Captive kestrels lay an average clutch of four eggs. Incubation follows, and ends when the first egg hatches. Thereafter, the parents feed their nestlings, and the period of parental care commences. The parental care period ceases when all young have fledged. Young birds was observed during the period of parental care. Since nestlings are inactive and are usually covered by the brooding female during the first 10 days after hatching, they were not observed before this time. The interval, from day 10 post-hatching until fledging, was termed the nestling period.

Test Schedule

Sampling was scheduled to document the activities of infected birds during every daylight hour. Cages to be observed were selected at random, but care was taken to distribute the observation sessions throughout the various periods of the day. Each day was separated into five observation periods: 6.00–9.00 hours (early morning), 9.00–12.00 hours (late morning), 12.00–15.00 hours (early afternoon), 15.00–18.00 hours (late afternoon), and 18.00 hours until dusk (evening). In early spring and late fall, the early morning and evening periods of observation were curtailed because of darkness. The above time intervals encompass the daily rhythms in behavioural patterns described by Smith *et al.* (1972) and Joly (1977). Sampling was not undertaken during periods of heavy rain, due to the reduced activity of the birds under such conditions. The behavioural repertoire of kestrels was divided into locomotory, perching, social, grooming, food handling, and breeding behaviours as well as miscellaneous of uncertain function.

BEHAVIOUR DESCRIPTIONS

Locomotory Behaviours

- *Flying*. In captivity, kestrels fly from perch to perch, to the nest box, or in tight horizontal loops.
- *Walking*. Birds walk on the ground or the platforms. This process involves alternate steps with an occasional hop.
- *Exercising*. Birds walks or perch while beating their wings, without becoming airborne (Sherrod 1983). This behaviour pattern, as well as mock attack on small, inanimate objects by young kestrels (Balgooyen 1976), was considered to be a form of exercising.
- *Perching behaviours*. Perching behaviours were divided into five categories on the basis of the bird's posture and alertness.
- *High perching*. Birds stand with both feet on the elevated structure, such as a wood or rope perch or on top of the next box, and appear to be keenly aware of their surroundings.
- *Floor perching*. Birds stand on the floor of the flight case and appear to the alert.
- *Nest perching*. Birds behave as above but are standing in the nest box (except when brooding on nest inspecting).
- *Resting*. Birds either lie on their bellies or support their weight on one leg with one foot tucked into the belly plumage. They appear to remain aware of their surroundings.

Social Behaviours

- *Aggressive displays*. Characteristically, aggressive displays involve a chase, with or without calls, and occasionally a jab with the talons as described by Cade (1955) and Mueller (1971).

- *Calling.* Adult kestrels possess a repertoire of three distinct calls. Vocalizations characteristically introduce or accompany other behavioural activities. When vocalization accompanied primary behavioural patterns, only the latter were recorded.

Grooming Behaviours

- *Preening.* Kestrels use their bill to nibble, stroke, and quiver their feather coats, modes in detail by Joly (1977) and Lefebvre and Joly (1982).
- *Scratching.* Kestrels use their feet to groom various body regions, particularly the head.
- *Comfort movements.* Comfort movements include stretching and bill-wiping, as described by Joly (1977).
- *Feather fluffing.* This activity consists of shaking the feather coat as described by Joly (1977).

Food Handling Behaviours

- *Feeding.* In the present study, birds were considered to be feeding only when they were actually tearing or ingesting prey.
- *Food transfer.* Food transfer between adults is usually followed by the presentation of food to the young by the female during the parental care period.

Breeding Behaviours

- *Courtship feeding.* The male presents food to the female. The behaviour consists of a complex of calls, postures, and coordinations, and serves to strengthen the pair bond (Balgooyen 1976).
- *Solicitation.* The female presents her back to the male, flicks her tail, and assumes a horizontal posture, thus offering a landing platform for the male during copulation.

- *Copulation.* Malting can occur regardless of whether the female was previously soliciting. This activity is well described by Willoughby and Cade (1964) and Balgooyen (1976).
- *Nestbox inspection.* This behaviour is most frequently performed by the male, and consists of foot scraping and loud chittering.
- *Incubation.* The process of sitting on eggs is generally performed by the female. The male assists the female, permitting her to leave the eggs to feed, preen, and stretch.
- *Brooding.* This behaviour refers to the act whereby the patterns shield their wings and body. Kestrels assume the brooding position by leaning forward and carefully stepping around the nestlings in a shuffling movement, thus placing their feet below the chicks.

Miscellaneous Behaviours

- *Tail flicking.* This behaviour consists of an up-and-down movement of the tail and is considered to be prelude to copulation.
- *Head bobbing.* This up-and-down movement of the head occurs when birds are intensely interested in whatever they are watching.
- *Gaping.* This behaviour can be compared to a short, forced yawn, and is actually repeated every few seconds.
- *Panting.* This behaviour resembles hard breathing with a partially open beak.

The present study concentrates on those behavioural processes that are affected by the *T. pseudospiralis* infections. Behavioural differences between infected and control birds that attributable to the infections are presented along with any interactions with the sex of the birds and/or the time of day. Where treatment effects are dependent on the time

Table 1. Courtship activity budget of *Trichinella pseudospiralis* infected kestrels. Data presented as percentage of time spent in each behaviour (mean ± standard error)

Flying	C		3.16 ± 0.44^a	*N*=162 (52)	Aggressive	C	F	EM $0.00 \pm$ o.00^b	*N*=17 (16)
(t, *p*=0.0306)	I		1.43 ± 0.28^b	*N*=104 (26)	displays			LM 0.00 ± 0.00^b	*N*=28 (26)
					S*T*TD,			EA 0.33 ± 0.33^b	*N*=20 (20)
					p=0.0040)			LA 0.00 ± 0.00^b	*N*=16 (16)
High perching	C	F	73.90 ± 2.80^a	*N*=81 (26)			M	Behaviour not observed	
(S*T, *p*=0.0028)		M	50.46 ± 3.44^b	*N*=81 (26)		I	F	EM 0.00 ± 0.00^b	*N*=13 (13)
	I	F	42.25 ± 5.46^b	N 52 (13)				LM 0.36 ± 0.19^a	*N*=14 (13)
		M	39.27 ± 4.81^b	*N*=52 (13)				EA 0.00 ± 0.00^b	*N*=12 (12)
								LA 0.00 ± 0.00^b	*N*=13 (12)
							M	Behaviour not observed	
Floor perching	C	F	0.37 ± 0.28^b	*N*= 81 (26)					
(S*T, *p*= 0.0001)		M	2.98 ± 1.34^b	*N*= 81(26)	Scratching	C		EM 0.78 ± 0.19^b	*N*=34 (32)
	I	F	35.71 ± 6.29^a	*N*= 52 (13)	T*TD,			LM 1.10 ± 0.25^b	*N*=56 (52)
		M	0.64 ± 0.35^b	*N*=52 (13)	*p*= 0.0276)			EA 0.58 ± 0.18^b	*N*=40 (40)
								LA 0.83 ± 0.29^b	*N*=32 (32)
						I		EM 0.38 ± 0.17^b	*N*=26 (26)
								LM 1.43 ± 0.24^a	*N*=28 (26)
								EA 0.28 ± 0.13^b	*N*=24 (24)
								LA 0.58 ± 0.21^b	*N*=26 (24)
Nest perching	C		1.20 ± 0.44^a	*N*=162 (52)	Copulation	C		0.56 ± 0.07^a	*N*= 162 (52)
(T, *p*= 0.0219)	I		0.05 ± 0.05^b	*N*= 104 (26)	T, *p*=0.0003)	I		0.22 ± 0.07^b	*N*=104 (26)
Sleeping		F	Behaviour not observed						

(S*T*TD), p=0.0201	C	M	EM 0.00 ± 0.00^{b}	N=17 (16)	Gaping (T*TD, p=0.0035)	C	EM 0.39 ± 0.35^{ab}	N=34 (32)
			LM 0.00 ± 0.00^{b}	N=28 (26)			LM $0.00 + 0.00^{b}$	N=56 (52)
			EA 0.00 ± 0.00^{b}	N=20 (20)			EA 0.00 ± 0.00^{b}	N=40 (40)
			LA $0.10 + 0.10^{ab}$	N=16 (16)			LA 0.00 ± 0.00^{b}	N=32 (32)
	I	F	EM 1.03 ± 1.03^{ab}	N= 13 (13)		I	EM 0.00 ± 0.00^{b}	N=26 (26)
			LM 0.00 ± 0.00^{b}	N= 14 (13)			LM 0.89 ± 0.66^{a}	N= 28 (26)
			EA $0.00 + 0.00^{b}$	N= 12 (12)			EA 0.00 ± 0.00^{b}	N=24 (26)
			LA $1.28 = 1.04^{a}$	N=13 (12)			LA 0.13 ± 0.09^{ab}	N=26 (24)
		M	EM 0.00 ± 0.00^{b}	N=13 (13)				
			LM 0.83 ± 0.83^{ab}	N=14 (13)				
			EA 0.00 ± 0.00^{b}	N=12 (12)				
			LA 0.00 ± 0.00^{b}	N=13 (12)				

p, probability obtained with GLM.

T, treatment; S, sex; TD, time of day; *, interactions.

C, control; I, infected; F, female; M, male.

EM, early morning; LM, late morning; EA, early afternoon; LA, late afternoon.

N, number of 15-min bird sessions of observation; (), number of individual birds observed.

Means with the same superscript are statistically indistinguishable using Duncan's multiple range test.

of day, only the means for the time period in which significant differences occurred are presented in the next.

Results

The birds initiated courtship behaviour as soon as they were paired on 10–13 April. Among control birds courtship extended, on average, over 32.27 ± 1.45 days (*N*=26), but lasted significantly longer among infected pairs (40 ±2.53 days, *N*=8); (*p*=0.0200). Incubation was first observed on 3 May and its duration was not significantly different (*p*=0.0200) for controls (30.32 ± 0.51 days, *N*=22) and infected birds (31 ± 1.7 days, *N*=3). The first eggs hatched on 2 June, marking the beginning of the period of parental care and the nestling period. This segment of the bird's life cycle continued an average of 27.90 ± 0.33 days (*N*=19) for control parents, and 29.00 ± 0.71 days for infected pairs (*N*=2). These differences were not statistically significant (*p*=0.400). The last young kestrel fledged on 30 July.

Courtship Period

Trichinella pseudospiralis infections influenced the flying activity of captive American kestrels. Infected kestrels were observed to fly less frequently than the control birds (T, *p*=0.0306) (Table 1). These effects were particularly noticeable among females. Infected females spent much of their time perching on the wire mesh floor. In contrast, control females were rarely seen to floor to floor perch (S*T, p=0.0001). Control females occupied high perches significantly more frequently than their mates, and significantly more frequently than the infected females and infected males (S*T, *p*=0.0028). Furthermore, control birds were observed to stand more frequently in the nest box than were infected birds (T, *p*=0.0219). No significant differences were found between infected and uninfected pairs with respect to their walking and exercising activities.

Gaping was more common among infected individual than among controls (T*TD, *p*=0.0035). This highly significant difference was apparently only during the late morning when gaping was characteristically most frequent. In addition, infected birds scratched more frequently during late morning than did the controls (T*TD p= 0.0276). Infected females slept more frequently during late afternoon than did female controls (S*T*TD, p=0.0201). The frequencies of resting, calling, preening, comfort movements, feather fluffing, feeding, and head bobbing were not found to be significantly different for control and infected pairs. Painting was not observed during the courtship period.

Aggressive displays were rarely witnessed during the courtship period, but statistical analysis revealed that infected females were more aggressive towards their mates, specifically during late morning, than were uninfected females (S*T*TD, p=0.0040). Infected birds flicked their tails more frequently during late morning (3.83 + 0.95, N= 56), than did the control pairs (4.10 + 1.04, N= 28) (T*TD, p= 0.0110). Due to the infrequency of this behaviour, it is not presented in Table 1. Copulation took place more frequently among controls than among infected pairs (T, p=0.0003). The frequencies of solicitation, nestbox inspection, and courtship feeding were not found to differ significantly for infected and control pairs.

Incubation Period

Seven infected females failed to produced complete clutches of six eggs and were thus excluded from observation during the incubation period. This reduced the sample size to six infected pairs. Since all of the uninfected females produced six eggs or more, the control sample size remained at 26 pairs.

The *T. pseudospiralis* infections had profound effects on the activity of female kestrels during the incubation period. In the late morning and late afternoon infected females spent most of their time perching on the floor of the flight pens. In contrast, during the same two periods, uninfected females were rarely observed to floor perch. These differences between infected and control birds were highly significant (S*T*TD, p=0.0101; Table 2). The frequencies of flying, walking, exercising, high perching, nest perching, resting, and sleeping did not differ significantly for controls and infected birds.

Panting was common during incubation, and was more frequently observed among infected birds than among controls during early afternoon, the warmest times of the day (T*TD, p = 0.0304). Furthermore, infected females were observed to feed more frequently than their uninfected counterparts (S*T, p = 0.0140). The frequencies of calling, preening, scratching, feather fluffing, head bobbing, comfort movements, tail flicking, and gaping were not significantly different for infected and control pairs.

Two breeding behaviours were affected by the *T. pseudospiralis* infection. Nestbox inspection was performed significantly more frequently by infected males than by the corresponding controls during late morning (S*T*TD, p= 0.0001; Table 2). Furthermore, control females were observed to incubate more frequently than were infected females (S*T, p= 0.0021). Concomitantly, infected males spent more than 30 per cent of their time incubating eggs during the late morning, whereas control males spent only 2–7 per cent of the time in this cavity. Aggressive displays, solicitation, and courtship feeding in controls and infected birds were statistically indistinguishable.

Parental Care Period

Pairs which failed to produce young, or lost their broods due to cannibalism, were eliminated from this group. This reduced the controls to 19 pairs, and the infected to three pairs. Since birds infected with *T. pseudospiralis* hatch fewer young, some of the most severely affected pairs may thus have been removed. No differences were found in the frequencies of flying and walking between infected and control birds. Exercising was not observed during this period in either infected or uninfected parents.

Regardless of the time of day or their sex, control birds occupied elevated perches more frequently than did infected individuals (T, p=0.0410). Furthermore, infected birds were not observed to scratch during the parental care period (N=24), whereas controls did (T, p=0.0482). During the early afternoon, panting was more frequent among infected birds than among the controls (T*TD, p=0.0090). Parenthetically, panting was not observed in the early morning and early evening when the temperatures were relatively cool. Tail flicking was rarely observed during this period, but analysis revealed that the infected males flicked their tail marginally more frequently during late morning (0.55 + 0.55, N=3) than did the corresponding controls (0, N=12) (S*T*TD, p=0.0457). Due to the infrequency of this behaviour, breeding behaviours were greatly reduced in frequency during the period of parental care. Incubation was succeeded by brooding, which did not differ between infected and control individuals. Other behaviours which did not appear to be affected by the infection were floor perching, nest perching, resting, sleeping, calling, preening, comfort movements, feather fluffing, feeding, food transfer, solicitation, copulation, gaping, and head bobbing. Aggressive displays were not observed during the parental care period.

During the parental care period, uninfected birds rearing infected young behaved differently from control birds rearing normal young. During late morning, gaping was observed in normal parents with infected young whereas normal parents with normal young were not observed to gape (T*TD, p = 0.006; Table 4). Similarly, male kestrels with infected young called significantly more frequently during late afternoon than the corresponding controls (S*T*TD, p=0.0259). No differences in the frequency of flying, walking, high perching, floor perching, resting, sleeping, preening, scratching, comfort movement, feather fluffing, feeding, food transfer, panting, tail flicking, and head bobbing were found between normal parents raising uninfected young and normal parents raising infected young. Exercising and aggressive displays were not observed in either of these parental groups.

Table 2. Incubation activity budget of *Trichinella pseudospiralis* infected kestrels. Data presented as percentage of time spent in each behaviour (mean ± standard error)

Floor perching	C	F	EM 0.36 ± 0.36^{b}	*N*=28 (23)	Nest inspection	C	F	EM 0.12 ± 0.12^{b}	*N*= 28 (13)
(S*T*TD, *p*=0.0101)			LM 0.00 ± 0.00^{b}	*N*=24 (23)	(S*T*TD)			LM 0.00 ± 0.00^{b}	*N*= 24 (23)
			EA 0.06 ± 0.06^{b}	*N*=28 (24)	*p*=0.0001)			EA 0.06 ± 0.06^{b}	*N*=28 (24)
			LA 0.43 ± 0.30^{b}	*N*=27 (23)				LA 0.00 ± 0.00^{b}	*N*=27 (23)
			EE 0.00 ± 0.00^{b}	*N*=12 (12)			M	EE 0.00 ± 0.00^{b}	*N*=12 (12)
		M	EM 0.83 ± 0.57^{b}	*N*=28 (23)				EM 0.00 ± 0.00^{b}	*N*=28 (23)
			LM 0.49 ± 0.36^{b}	*N*=24 (23)				LM 0.00 ± 0.00^{b}	*N*=24 (23)
			EA 0.30 ± 0.21^{b}	*N*=28 (24)				EA 0.12 ± 0.12^{b}	*N*=28 (24)
			LA 1.60 ± 1.48^{b}	*N*=26 (23)				LA 0.00 ± 0.00^{b}	*N*=26 (23)
			EE 0.24 ± 0.24^{b}	*N*=14 (13)				EE 0.00 ± 0.00^{b}	*N*=14 (13)
	I	F	EM 17.22 ± 17.23^{b}	*N*=3 (3)		I	F	Behaviour not observed	
			LM 60.00 ± 30.56^{a}	*N*=3 (3)			M	EM 0.00 ± 0.00^{b}	*N*=3 (3)
			EA 0.00 ± 0.00^{b}	*N*=6 (4)				LM 3.33 ± 2.55^{a}	*N*=3 (3)
			LA 64.72 ± 20.50^{a}	*N*=6 (4)				EA 0.00 ± 0.00^{b}	*N*=6 (4)
			EE 5.55 ± 5.56^{b}	*N*=3 (3)				LA 0.00 ± 0.00^{b}	*N*=6 (4)
		M	EM 0.00 ± 0.00^{b}	*N*=3 (3)				EE 0.00 ± 0.00^{b}	*N*= 3 (3)
			LM 5.00 ± 5.00^{b}	*N*=3 (3)					
			EA 1.67 ± 1.67^{b}	*N*=6 (4)					
			LA 0.00 ± 0.00^{b}	*N*=6 (4)					
			EE 0.00 ± 0.00^{b}	*N*=3 (3)					

Feeding (S*T, p=0.0140)	C	F	1.81 ± 0.67[a]	N=119 (26)	Panting (T*TD*, p=0.0304)	C	EM 0.00 ± 0.00[b]	N 56 (46)
		M	5.07 ±1.44[a]	N= 120 (26)			LM 0.83 ± 0.46[b]	N=48 (48)
	I	F	3.02 ±1.50[b]	N=21 (6)			EA 0.03 ± 0.03[b]	N=56 (48)
		M	3.75 ± 3.40[ab]	N=21 (6)			LA 0.85 ± 0.61[b]	N=53 (46)
Incubation (S*T, p= 0.0021)	C	F	63 ± 3.47[a]	N=119 (26)		I	EE 0.00 ± 0.00[b]	N=26 (4)
		M	3.88 ± 1.26[a]	N=120 (26)			EM 0.00 ± 0.00[b]	N=6 (6)
	I	F	30.85 ± 10.42[b]	N=21 (6)			LM 0.00 ± 0.00[b]	N=6 (6)
		M	9.53 ± 6.57[a]	N=21 (6)			EA 5.56 ± 3.63[a]	N=12 (8)
							LA 15.56 ± 10.51[ab]	N=12 (8)
							EE 0.00 ± 0.00[b]	N=6 (6)

p, probability obtained with GLM.

T, treatment; S, sex; TD, time of day; *, interactions.

C, control; I, infected; F, female; M, male.

EM, early morning; LM late morning; EA, early afternoon; LA, late afternoon.

N, number of 15-min bird sessions of observation; (), number of individual birds observed.

Means with the same superscript are statistically indistinguishable using Duncan's multiple range test.

Table 3. Parental care activity budget of *Trichinella pseudospiralis*-infected kestrels. Data presented as percentage of time spent in each behaviour (mean ± standard error)

High perching (28)	C		59.15 ± 3.31^{a}	*N* = 1 3 6
(T, *p*= 0.0410)	I		49.94 ± 7.58^{b}	*N*= 24 (6)
Scratching (28)	C		0.37 ± 0.08^{a}	*N* = 1 3 6
(T, *p*= 0.0482)	I		0.00 ± 0.00^{b}	*N*= 24 (6)
Panting (T*TD, *p*= 0.0090)	C	EM	$0.00 + 0.00^{d}$	*N*=26 (20)
		LM	5.14 ± 3.41^{cd}	*N*=24 (24)
		EA	5.39 ± 4.09^{cd}	*N*=26 (22)
		LA	1.50 ± 1.04^{cd}	*N*=30 (26)
		EE	0.00 ± 0.00^{d}	*N*=30 (22)
	I	EM	0.00 ± 0.00^{d}	*N*=2 (2)
		LM	14.73 ± 11.71^{bc}	*N*=6 (4)
		EA	27.23 ± 10.97^{a}	*N*=6 (6)
		LA	20.29 ± 12.69^{ab}	*N*=6 (6)
		EE	$0.00 + 0.00^{d}$	*N*=4 (4)

p, probability obtained with GLM.

T, treatment; S, sex; TD, time of day; *, interactions.

C, control; I, infected; F, female; M, male.

EM, early morning; LM, late morning; EA, early afternoon; LA, later afternoon.

N, number of 15-min bird sessions of observations; (), number of individual birds observed.

Means with the same superscript are statistically indistinguishable using Duncan's multiple range test.

During the parental care period, copulation was observed significantly more frequently among control parents caring for infected young, specifically during the early afternoon and evening (T*TD, *p*, 0.0.0165). Control parents with normal nestlings were observed to copulate only during early morning and significantly less frequently than the preceding group (T*TD, *p*=0.0165). In addition, uninfected females rearing normal young perched more frequently in the nest than did uninfected females rearing infected young (S*T, p=0.0163). Nest perching in uninfected adult males caring for normal nestlings did not differ from that of their counterparts which cared for infected nestlings. Brooding and solicitation by uninfected parents did not appear to be affected by the presence of infected nestlings. Nestbox inspection was not observed during this period.

Nestling Period

Uninfected male nestlings reared by infected parents exercised more frequently than their counterparts reared by control parents (S*T, p=0.0231). Such differences were not significant among uninfected female nestlings cared for by infected parents. The frequency of exercising among infected male nestlings did not differ significantly from that of their uninfected counterparts when both were reared by normal parents. The frequency of exercise among infected female nestlings reared by normal parents did not differ from their uninfected counterparts. Uninfected young raised by infected parents were observed to perch only marginally less frequently than were uninfected nestlings reared by normal parents (T, p=0.0487). Furthermore, the frequency of perching among infected nestlings of both sexes did not differ significantly from that among their uninfected counterparts when both were reared by normal parents. Calling was affected by the infection. The uninfected young of infected parents were observed to call significantly more frequently during late afternoon than were infected young of normal parents and infected young raised by normal parents (T*TD, p=0.0003). There was no difference in aggressive displays for any of the nestling groups.

The frequency of panting among female nestlings was affected by whether they were reared by infected or normal parents. During the early afternoon uninfected female nestlings raised by infected parents panted more frequently than uninfected female nestlings raised by normal parents (S*T*TD, p=0.0002). Infected female nestlings raised by normal parents panted with a frequency of 60.00 ± 30.56 per cent (N=3). This differs significantly from their uninfected

Table 4. Parental care activity budget of uninfected kestrels raising *Trichinella pseudospiralis*-infected young, Data presented as percentage of time spent in each behaviour (mean ± standard of error)

Behaviour	Group	Sex	Mean ± SE	N	Behaviour	Group	Sex	Mean ± SE	N
Nest perching	Pyc	F	12.68 ± 4.62^{a}	N= 46 (10)	Calling	Pyc	F	EM $0.00 + 0.00^{c}$	N=9 (7)
(S* T, p= 0.0163		M	0.33 ± 0.16^{b}	N=46 (10)	(S*T*TD),			LM 0.21 ± 0.21^{c}	N= 8 (8)
	Pyi	F	0.23 ± 0.17^{b}	N=22 (5)	p= 0.0259)			EA 0.37 ± 0.37^{c}	N=9 (8)
		M	0.30 ± 0.18^{b}	N=22 (5)				LA 0.67 ± 0.67^{c}	N= 10 (9)
								EE 0.00 ± 0.00^{c}	N=10 (9)
Copulation	Pyc		EM 0.19 ± 0.13^{b}	N=18 (14)			M	EM 1.30 ± 0.77^{abc}	N=9 (7)
(T*TD, p= 0.0165)			LM $0.00 + 0.00^{b}$	N=16 (16)				LM 1.46 ± 1.46^{bc}	N= 8 (8)
	C	F	EA 0.00 ± 0.00^{b}	N=18 (16)				EA 0.00 ± 0.00	N=9 (9)
			LA 0.00 ± 0.00^{b}	N=20 (18)				LA 0.00 ± 0.00^{c}	N= 10 (9)
			EE 0.00 ± 0.00^{b}	N=20 (18)				EE 0.33 ± 0.22^{bc}	N=10 (9)
	Pyi	M	EM 0.00 ± 0.00^{b}	N=8 (6)		Pyi	F	Behaviour not observed	
			LM $0.00 + 0.00^{b}$	N=8 (8)			M	EM 0.00 ± 0.00^{c}	N=4 (3)
			EA 0.42 ± 0.27^{a}	N=8 (6)				LM 0.750 ± 6.96^{ab}	N=4 (4)
			LA 0.00 ± 0.00^{b}	N=10 (8)				EA 0.00 ± 0.00^{c}	N= 4 (3)
			EE 0.33 ± 0.22^{a}	N=10 (8)				LA 10.00 ± 7.66^{a}	N= 5 (4)
								EE 3.00 ± 3.00^{bc}	N=5 (4)
Gaping	Pyc		Behaviour observed						not

(T*TD, p= 0.0006)	Pyi	EM 0.00 ± 0.00^{b}	N=8 (6)
		LM 0.42 ± 0.27^{a}	N=8 (8)
		EA 0.00 ± 0.00^{b}	N=8 (6)
		LA $0.00 + 0.00^{b}$	N=10 (8)
		EE $0.00 + 0.00^{b}$	N= 10 (8)

p, probability obtained with GLM.

T, treatment; S, sex; TD, time of day; *, interactions.

Pyc, parents with uninfected young; Pyi, parents with infected young; I, infected; F, female; M, male.

EM, early morning; LM, late morning; EA, early afternoon; LA, late afternoon.

N, number of 15-min bird sessions of observation; (), number of individual birds observed.

Means with the same superscript and statistically indistinguishable using Duncan's multiple range test.

counterparts which were not observed to pant (*N*=13), (S*T*TD, *p*=0.0002). Male nestlings did not exhibit such differences. The frequencies of preening, scratching, comfort movement, feather fluffing, tail flicking, head bobbing, feeding, and gaping were not affected by the parasitic infection.

The frequency of sleeping among nestling was also affected by whether they were reared by infected or control parents. Uninfected young reared by infected parents slept significantly more frequently than their counterparts reared by normal parents, and also more frequently than the infected nestlings (T*TD, *p*=0.0003). This difference was apparent only during the late morning observation period. There was not difference in resting for any of the nestling groups.

The mean total larvae recovered was 310560.92 ± 76024.69 for the 13 infected adult females and 116099.08 + 23.745.17 (*N*=13) for the adult males. These represent, in our experience, relatively moderate infections. None of the infected birds died nor manifested clinical disease.

Discussion

Experimental *T. pseudospiralis* infections altered the behavioural processes of the adult kestrel host. Not all of these effects are seen in both sexes, during all periods of the host' life cycle, or during all times of the day. Infected birds exhibited effects that were consistent with the pathological tissue changes described from avian hosts infected with this parasite, as well as with the well-documented symptoms accompanying *T. spiralis* infections in the human host. Thus, during the courtship period, infected birds of both sexes were observed to fly less frequently than uninfected controls. This effect may be attributable to muscle damage. In addition, infected kestrels also gaped more frequently than the controls, specifically during late morning. Conceivably, gaping is a manifestation of lung damage inflicted by migrating larvae. Larval migration is a transitory phenomenon and characterizes that early stages of infection. In the present study, this phase fell within the courtship period of the birds. Since the observed differences in the frequency of gaping between infected and control birds were highly significant and since this behaviour did not show significant differences during subsequent periods of observation, they are considered to be transitory manifestations of infection. Similarly, infected birds of both sexes scratched more frequently (late morning) than did the controls during the courtship period. These differences were highly significant during the early stages of the infection, and at best only marginally significant during the subsequent periods. Conceivably, scratching is a responses to the itching sensation described by Gould (1945) in cases of human trichinellosis.

The data suggest that *T. pseudospiralis* may create some thermal stress. On hot days, kestrels pant to maintain thermal homeostasis. During the incubation and parental care periods, particularly in the early afternoon, infected birds of both sexes were observed to pant more frequently than the controls. The pathological basis for this effect is obscure, but may be related to the pyrogenic response to the invading parasites. The consequences of infection seem to be much more severe in female than in male birds. Thus, during the courtship, period, infected females were observed to sleep in the daytime, especially during late afternoon, and to feed more during the incubation period than did uninfected controls. Both activities may be associated with the demands of tissue repair.

During the courtship period, infected individuals flicked their tail more frequently (late morning) than their uninfected counterparts. Since tail flicking may be considered to be a prelude to copulation and since copulation among infected pairs was significantly less frequent than among controls, the elevated frequency of tail flicking among infected males may indicate a lowered threshold for this activity. During the courtship period, infected females

Table 5. Nestling activity budget of *Trichinella pseudospiralis*-infected young. Data presented as percentage of time spent in each behaviour (mean ± standard error)

Behaviour	Young	Parent	Time	Mean ± SE	N
Perching (T, p, =0.0487)	NY	(NP)		52.93 ± 4.30^{a}	N=78 (20)
	IY	(NP)		51.91 ± 5.14^{ab}	N=56 (18)
	NY	(IP)		36.25 ± 8.46^{b}	24 (7)
Calling (T*TD, p=0.0003	NY	(NP)	EM	0.00 ± 0.00^{b}	N=13 (13)
			LM	1.78 ± 0.93^{ab}	N= 15 (15)
			EA	0.00 ± 0.00^{b}	N=18 (6)
			LA	0.00 ± 0.00^{b}	N= 16 (16)
			EE	0.21 ± 0.21^{b}	N=16 (16)
	IY	(NP)	EM	0.13 ± 0.13^{b}	N = 13 (13)
			LM	0.00 ± 0.00^{b}	N=11 (11)
			EA	0.00 ± 0.00^{b}	N=7 (7)
			LA	0.00 ± 0.00^{b}	N=16 (16)
			EE	1.67 ± 1.67^{b}	N=9 (9)
	NY	(IP)	EM	$0.00^{b} \pm 0.00^{b}$	N=2 (2)
			LM	$0.00^{b} \pm 0.00^{b}$	N=5 (5)
			EA	$0.00^{b} \pm 0.00^{b}$	N=5 (5)
			LA	2.78 ± 1.36^{a}	N=9(8)
			EE	$0.00^{b} \pm 0.00^{b}$	N=3 (3)
Sleeping (T*TD, p=0.0003)	NY	(NP)	EM	0.00 ± 0.00^{b}	N=13 (13)
			LM	0.00 ± 0.00^{b}	N=15 (15)
			EA	23.43 ± 8.99^{b}	N=18 (16)
			LA	1.35 ± 1.35^{b}	N=16 (16)
			EE	$0.00^{b} \pm 0.00^{b}$	N=16 (16)

Behaviour	Young	Parent	Sex	Time	Mean ± SE	N
Exercising (S*T, p= 0.0231)	NY	(NP)	F		0.65 ± 0.27^{b}	N=59 (14)
			M		0.09 ± 0.09^{b}	N= 19 (b)
	IY	(NP)	F		0.43 ± 0.17^{ab}	N=27 (8)
			M		0.23 ± 0.14^{b}	N=29 (10)
	NY	(IP)	F		0.00 ± 0.00^{v}	N=15 (4)
			M		0.93 ± 0.40^{a}	N=9 (3)
Panting (S*T*TD, p=0.0002)	NY	(NP)	F	EM	0.00 ± 0.00^{c}	N=10 (10)
				LM	7.12 ± 3.90^{cde}	N=11 (11)
				EA	0.00 ± 0.00^{e}	N=13 (11)
				LA	13.21 ± 8.10^{cde}	N=13 (13)
				EE	0.00 ± 0.00^{e}	N=12 (13)
			M	EM	0.00 ± 0.00^{e}	N=3 (3)
				LM	0.00 ± 0.00^{e}	N=4 (4)
				EA	20.00 ± 20.00^{cde}	N=5 (5)
				LA	31.11 ± 24.76^{abcd}	N=3 (3)
				EE	0.00 ± 0.00^{e}	N=4 (3)
	IY	(IP)	F	EM	0.00 ± 0.00^{e}	N=7 (7)
				LM	6.39 ± 6.39^{de}	N=6 (6)
				EA	60.00 ± 30.56^{abc}	N=3 (3)
				LA	0.00 ± 0.00^{e}	N=7 (7)
				EE	0.00 ± 0.00^{e}	N=4 (4)

IY	(NP)	EM	2.56 ± 2.57^{b}	N=13 (13)
		LM	0.00 ± 0.00^{b}	N=11 (11)
		EA	12.38 ± 12.39^{b}	N=7 (7)
		LA	6.04 ± 2.85^{b}	N=16 (16)
		EE	8.33 ± 4.31^{b}	N=9 (9)
NY	(IP)	EM	0.00 ± 0.00^{b}	N=2 (2)
		LM	30.00 ± 8.62^{a}	N=5 (5)
		EA	3.33 ± 3.33^{b}	N=5 (5)
		LA	0.00 ± 0.00^{b}	N=9 (8)
		EE	0.00 ± 0.00^{b}	N=3 (3)

		M	EM	0.00 ± 0.00^{e}	N=6 (6)
			LM	6.67 ± 6.67^{e}	N=5 (5)
			EA	11.67 ± 11.67^{cde}	N=4 (4)
			LA	0.74 ± 0.74^{a}	N=9 (9)
NY	(IP)	F	EM	0.00 ± 0.00^{e}	N=1 (1)
			LM	0.00 ± 0.00^{e}	N=3 (3)
			EA	100.0 ± 0.00^{a}	N=3 (3)
			LA	60.67 ± 22.20^{ab}	N=5 (4)
			EE	0.00 ± 0.00^{e}	N=3 (3)
		M	EM	0.00 ± 0.00^{e}	N=1 (1)
			LM	$20.83 + 20.83^{bcdc}$	N=2 (2)
			EA	0.00 ± 0.00^{e}	N=2 (2)
			LA	0.00 ± 0.00^{e}	N=4 (4)

p, probability obtained with GLM.

T, treatment; S, sex; TD, time of day; *, interactions.

NY (NP), normal young raised by normal parents; IY (NP), infected young raised by normal parents; NY (IP), normal young raised by infected parents; F, female; M, male.

N, number of 15-min bird sessions of observation; (), number of individual birds observed.

Means with the same superscript are statistically indistinguishable using Duncan's multiple range test.

exhibited an increase in the frequency of aggressive displays (late morning). This suggests that they may be less receptive to mating than uninfected females. A lack of receptivity has been documented from female mice infected with *T. spiralis*. Edwards and Barnard (1987) report that infected female mice avoid, or act aggressively towards males that attempt copulation.

Since infected birds which failed to produce complete clutches or to produce young at all (presumably the most severely affected individuals) had been removed from the experimental group, the observed changes in the behaviour of the remaining infected pairs are conceivably an underestimate of the effects of *T. pseudospiralis* infection.

Parasite-associated changes in kestrel behaviour extend to other activities as well. Infected birds of both sexes were seen to perch less frequently in their elevated nest boxes during the courtship period than did the corresponding controls. Infected females spent a greater amount of time on the floor of the flight cage both during the courtship and incubation period (late morning and late afternoon). Similarly, high perching was significantly more frequently among control females during the courtship period, then it was among their infected counterparts. This was also true for female and male controls during the period of parental care. Indeed, infected females tend to lay their eggs on the floor rather than in the nest box. Such eggs usually break or are abandoned by the females. Furthermore, infected females were observed to incubate significantly less frequently than their uninfected counterparts. It is conceivable that the elevated prevalence of nest inspection during the incubation period exhibited by their mates during late morning constitutes a response compensatory to the lack of care shown by the infected females. Indeed, at certain times of the day, the males of infected pairs spent more time incubating then the males of control pairs. Infected pairs produce significantly fewer offspring than their uninfected counterparts.

The altered behaviour of infected parents may affect the behaviour of their offspring. Uninfected nestlings reared by infected parents manifested signs of restlessness. Thus, both male and female nestlings called more frequently during late afternoon, and young, uninfected males reared by infected parents exercised more than their counterparts reared by normal birds. Calling by nestling kestrels is a sign of hunger or stress. Conceivably, nestlings reared by infected parents suffered some degree of nutritional stress. However, since female nestlings are generally more aggressive and tend to be the first to get fed, this stress may be more severe among young males. Indeed, nestling females were observed to feed more frequently, whether infected or uninfected. Hunger will increase the activity of small mammals. Conceivably, it may have similar effects on this raptor species. Following a period of high activity, young kestrels usually become quiescent. Uninfected nestlings raised by infected parents were observed to sleep more during late morning and perch less than did their counterparts raised by control birds. This increase in the time spent sleeping may be the result of fatigue from begging for food. Also, possible differences in the ability to cope with heat stress, as manifested by abnormally frequent panting among the uninfected female nestlings raised by infected parents (early afternoon), may be due to hunger. What may be nutritionally mediated effects of maternal infection on the behaviour of their offspring have been reported by maternal infection on the behaviour of their offspring have been reported by Rau (1985) from mice harbouring *T. spiralis*.

Although the effects of the parental infection on the behaviour of the nestlings are pronounced, the infection of the young themselves had little impact. This may be due to the fact that the infection in the nestling falls within the enteric phase of the parasite's life cycle, and larvae are produced and disseminated into the tissues only when nestlings are about to fledge. According to Meerovitch *et al.* (1982), larvae of *T. pseudospiralis* are first seen in the musculature of kestrels on day 20 post-infection. In *T. spiralis* infection of mice behavioural changes accompany the release of migrating larvae. No behavioural changes were attributable to the enteric phase of the infection. It is conceivable that, as with *T. spiralis*, this phase of *T. pseudospiralis* infection is relatively benign.

The presence of infected nestlings also alters the behaviour of the normal parents. Females rearing infected nestlings spent less time in the nest with these young, and both such parents gaped more frequently during late morning than did parents rearing uninfected nestlings. In addition, the male parents rearing infected young called more during late afternoon than their counterparts. Such parents copulated during early afternoon and evening at a time in their reproductive cycle when their uninfected counterparts had largely abandoned this activity. This mechanisms by which these behavioural changes are induced in the uninfected bird are still obscure.

Trichinella pseudospiralis infections interfere with a wide variety of behavioural processes in the kestrel host. The behavioural changes induced by *T. pseudospiralis* in kestrels resemble those induced in mice by *T. spiralis*. Both host species experience significant reductions in mobility as a result of infection with their respective parasites. However, mice infected with *T. pseudospiralis* exhibit quite different behavioural modifications. Instead of reduction in activity, *T. pseudospiralis*-infected mice, particularly heavily infected individuals, demonstrated a mild increase in activity. *Both T. spiralis* in mice and *T. pseudospiralis* in kestrels elicit severe inflammatory responses in the musculature, *T. pseudospiralis* in mice does not. Indeed, *T. pseudospiralis* has an anti-inflammatory action in the latter host. Conceivably, many of the behavioural changes manifested during the chronic phase of *T. pseudospiralis* infections in kestrels, as well as *T. spiralis* and *T. pseudospiralis* infections in mice, are attributable, at least in part, to the character and severity of the host responses elicited.

Chapter—7
Trichuris

Worms of the order Trichinellida are characterized by the possession of an enormously elongated capillary-like oesophagus, whose internal cavity is so small that it is difficult to locate even in sections. The oesophagus is embedded in a long column of single cells (stichocytes) which may serve as oesophageal glands, the whole structure being referred to as a *stichosome.* Ultrastructure studies show that the oesophagus is made up of a single tubular cell separated from the stichocytes by a membranous sheath. The cuticle is also unusual in possessing a 'bacillary band' consisting of bacillary cells which penetrate the cuticle over about one-third of its circumference. Ultrastructure studies suggest that these cells may possibly have an excretory role.

The anus is at the extreme posterior tip, and not just in front of the hind end, so that there is no tail. The male has one spicule or none at all; if present, it is enclosed in a protrusible membranous sheath.

The major families of order Trichinellida are diagnosed as follows:

Trichuridae

The 'whip worms'. Oviparous. Adults in intestine. Males with protrusible spiny sheath and spicule (*e.g. Trichuris trichiura*).

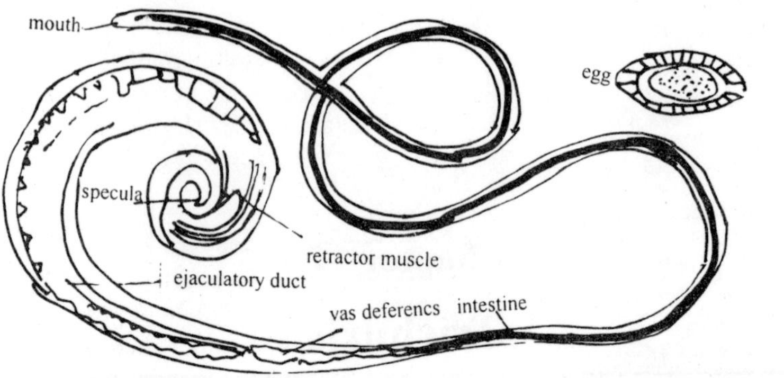

Fig. 7.1 : *Trichuris trichiura*—morphology of male and egg.

Trichinellidae

Viviparous. Male with no spicule or spicule or spicule sheath. Larvae in tissue; adult in intestine (only one species: *Trichinella spiralis*).

Trichosomoididae

Ovoviviparous. Adults in urinary bladder of rodents. Male minute and parasitic in uterus of female worm (*e.g. Trichosomoides crassicauda*).

FAMILY TRICHURIDAE

Genus: Trichuris

These are the so-called 'whip worms', a term derived from the whip-like form of the body.

The correct generic name for members of this group is *Trichuris*—a term which given a misleading description, for it literally means 'hair tail'. The name was applied before it was realized that the thin thread-like part which looked like the tail was, in fact, the head. A more apt name, then, is *Trichocephalus*, which was later applied, but the earlier name *Trichuris* has zoological priority.

The best known species are as follows:

Trichuris muris : rodents.

T. ovis: sheep, goats, cattle and other herbivores.

T. trichiura: man, and monkeys.

T. suis: pigs.

General Morphology

This does not vary much. The thread-like oesophageal region occupies about two-thirds the length of the body. The mouth has no lips. The vulva is at the junction of the thread like and thickened of the body. The eggs have a characteristic barrel shape with an opercular plug at each end.

Life Cycle

The life cycle is similar in all species. Eggs require a worm moist environment for embryonation (such as that provided by rain-soaked soil), but once embryonated are exceptionally resistant to environmental conditions. Embryonated eggs hatch near the caecum into which the larvae migrate. Possibly related to scarcity of food materials in this site, development is slow, and 70-90 days are required for maturation. Heavy infections in humans cause abdominal discomfort, anemia and eosinophilia.

As the eggs of both *Ascaris* and *Trichuris* require the same conditions for embryonation (high humidity and access to oxygen) these two species frequently occur together in humans. This is well demonstrated in infection figures in American school children for these two species. The pathology of trichuriasis is reviewed by Beck and Beverley-Burton (1968).

Genus: *Capillaria* (included in a separated family Capillariidae by some workers).

In this genus, the characteristic division into a narrow anterior region and a broader posterior region is not so marked. Worms are largely parasitic in a number of warmblooded animals, particularly birds and small carnivorous mammals, but they also occur in fish, amphibia and reptiles.

Life History

The life history varies rather unusually with species. In *C. obsignata* (=*columbae*) (pigeons) and *C. aerophila* (foxes) development is similar to *Trichuris* with direct embryonation of the eggs outside the host and reinfection by swallowing embryonated eggs. In *C. hepatica* (rodents), the adults occur in the liver and released eggs accumulate on the surface of the liver. As the eggs require oxygen for normal embryonation, ingestion of an infected liver by another animal will not lead to an infection, but eggs subsequently passed undigested into the faeces, when exposed to air and moisture, become infective on re-ingestion.

Table 1. Prevalence of intestinal parasites in stool specimens from 631 Cherokee Indian elementary school children in the U.S.A.

Parasite	Specimens positive	
	Number	*Per cent*
Helminths:		
Ascaris lumbricoides	312	49
Trichuris trichiura	240	38
Hookworm	19	3
Trichostrongylus species	1	<1
Enterobius vermicularis	14	2
Protozoa:		
Entamoeba histolytica	67	11
Entamoeba hartmanni	220	35
Entamoeba coli	251	40
Endolimax nana	289	46
Iodamoeba butschlii	34	5
Giardia lamblia	59	9
Dientamoeba fragilis	68	11
Trichomonas hominis	72	11
Chilomastix mesnili	19	3
Unidentified protozoa	7	1
1 or more parasites	579	92
No parasites found	52	8

The capillarid worms *C. longicollis* and *C. annulata* make use of earthworms as intermediate hosts. Embryonated eggs are not infective if swallowed directly by the definitive host, but if ingested by an earthworm become infective within twenty-four hours. The larvae are released within the earthworm and enter the connective tissue of the body cavity; the bird hosts become infected by eating earthworms.

The life history and pathology of species of *Capillaria* have been reviewed by Beck and Beverley-Burton (1968).

Family Trichinellidae

This family contains only a single genus *Trichinella* (ayn. *Trichina*) and a single species *Trichinella spiralis*, the 'trichina'—an ubiquitous parasite which infects a wide spectrum of mammalian hosts. The chief host are primarily rodents (especially rats), pigs man and carnivores but more than 100 species have been reported as hosts. The parasite causes trichinellosis (trichinelliasis, trichinosis) in man.

Trichinellosis has been known for centuries in Eastern Europe being especially prevalent in Poland, west and south-east Russia, Rumania, Hungary and Czechoslovakia. It is rare in Austria, Finland, Denmark, Holland, France and Belgium but known to occur in North America. There are occasional local outbreaks in the U.K. Until recently, it was generally accepted that the disease was limited to temperate and arctic regions. It is now known to occur in the tropics also and has been reported from Algeria, the Middle Easts China, Hawaii, Africa and (rarely) South America. It has also been found in New Zealand but (to date) appears to be absent from Australia. The large intake of migrants into Australia may result in its ultimate spread to that country also.

An enormous literature on trichinellosis exists and only a general account is given below. Some excellent reviews and conference reports on this topic are available such as those of Lupascu (1970), Beck and Beverley-Burton (1968), Kozar (1967) and Larsh (1963, 1968).

Morphology

Both sexes are small, the male (1.4-1.6 mm) being smaller and more slender than the female (3-4 mm). Spicules are absent in the male, but the cloaca is everted during copulation and its opening flanked by two papillae. The female genital system is single. The ultrastructure has been described by the Beckett and Boothroyd (1961) and Simionescu *et al.* (1969).

Life History

Infection is brought about by a host eating meat containing encapsulated larvae. These larvae are released in the duodenum and may be found there as early as one hour after the initial infection.

Deeply embedded in the mucosa, larvae, feed, grow and rapidly become sexually mature. Female worms have been found inseminated at 36 hours which suggested that copulation first takes place at about 30-35 hours. There is evidence from *in vitro* studies that a chemically mediated sexual attraction operates between male and female worms. The effectiveness of this attraction mechanism (*in vivo*) is remarkable and even mice inoculated orally with only *two Trichinella* larvae have been shown to develop patent infections. The females are ovoviviporous and up to 1500 larvae may be produced by a single worm.

Released larvae penetrate the mucosa and are carried to the voluntary muscles, especially those of the diaphragm, jaws, tongue, larynx and eye. Since larvae may be detected both in peritoneal washings and in the blood stream on the sixth day post-infection it is clear that both peritoneal and lymph vessel-blood stream migration routes to the muscles may be followed by the larvae. Controversy exists, however, on the relative importance of these routes.

According to Shanta and Meerovitch (1967*b*) the development phase in the muscle is as follows: larvae commence penetrating the diaphragm muscle on the sixth day, piercing the sarcolemma of the fibres and initially lying extended. On the 17th day, they begin to coil and on the 21st day the formation of a cyst wall can be detected and encapsulation follows: the larvae becoming resistant to pepsin digestion at 19 days. Calcification of cysts may occur as early as five months, but usually begins after 6-18 months.

Larvae may remain visible for several years within the cysts but do not develop further. Even at this stage, however, the genitalia primordia are well developed although differences between males and females cannot be detected. When infected muscles are eaten by the definitive host, sexual maturation in the intestinal phase rapidly occurs, as explained above.

Number of Moults

Controversy exists on the number of moults undertaken by *T. spiralis* during its life cycle—especially in the intestinal phase—one to four moults being reported by different workers. According to Shanta and Meerovitch (1967*a*, *b*) only *two* moults occur, both during the intestinal phase and none occurs in the muscle phase. They put forward two possible hypotheses either:

(*a*) during evolution the last two moults (*i.e.* the third and fourth) disappeared, so that the infective larva (penetrating the muscle) may be considered to be a third-stage larva; *or*

(*b*) young (*i.e.* newly hatched) larvae do not moult at all, in which case the encapsulated larva in muscle would be in first stage. Since, when taken into the host and pigs become infected by eating infected pork scraps or occasionally rats which penetrate into their stalls. Both rats and pigs tolerate infections well. Man becomes infected by eating imperfectly cooked pork and, as mentioned, may reinfect pigs by feeding back scraps of infected pork in pig food. Sausages are a dangerous source of parasite, as a small piece of pork after mincing may be distributed amongst a number of sausages, although, as factories pool pork from many animals, considerable dilution may take place.

The realization that 'strains' of *T. spiralis* exist arose from investigations largely by Nelson and his co-workers of an outbreak of trichinellosis in Kenya. Other strains have been reported from India.

The East African strain was found to have alow infectivity to both pigs and rats and a high infectivity but a low pathogenicity to man. In the feral (=wild) cycle in Africa and also in the Arctic, rats are not involved and transmission is largely in and by carnivores, with the bush-pig playing a subsidiary role. It is thus probably only in urban areas (*i.e.* in Europe and America) that rats play a major role in the life cycle. As with many other strains of parasites, however, it must be emphasized that laboratory passage may greatly alter characteristics such as infectivity and pathogenicity.

Incidence in Man

It is difficult to obtain accurate figures for the incidence of trichinellosis in man as the disease often remains undetected (in light infections) during life. Serological tests (such as bentonite or charcoal flocculation, latex agglutination or skin tests) are available, but a number of surveys have also been carried out by examining the diaphragms of cadavers. In the U.S.A. the incidence in cadavers over the period 1936-41 reached the astonishing figure of 16 per cent. A more recent survey for the period 1966-68 showed a very marked reduction, the incidence in 5000 human diaphragms being 4.2 per cent. In Poland the incidence in man is only about 4 per cent but the morbidity is high. World figures fro the incidence of trichinellosis have been reviewed in detail by Lupascu (1970).

Methods of Detection

Two methods are used for detecting the organisms in muscle samples (*a*) artificial digestion (by pepsin) or (*b*) by means of a 'trichinoscope'—a simple device which compresses a sample of muscle between glass plates in a manner which makes it semi-transparent. According to Zimmerman *et. al.* (1968) the digestion method is the efficient.

Control

It is beyond the scope of this book to discuss the detailed prophylaxis of trichinelliasis; much of this is clear from considerations of the life cycle. It is evident, however, that storage of pig meat in deep freeze units or adequate cooking to high temperatures would serve as major control measures. For this reason, the effect of temperature on *T. spiralis* is of special interest.

A temperature of 55°C kills the larvae within 15 minutes in pig meat. Periods of freezing to kill larvae in pig meat at different temperatures are: 30 minutes at –40°C; 24 hours at –25°C; 3 days at –20°C and 6 days at –15°C (Podhajecky, 1964*a*). The U.S. Bureau of Animal Industry makes use of this knowledge and requires uncooked pork to hang at –15°C for twenty days before distribution.

Immunity

Animals including man, rapidly develop an immunity to *Trichinella*. Antigens have been found in the blood as early as twenty-four hours after the initial infection. Although complete immunity is not found, a high degree of protection is developed, and worms in immunized hosts show retarded development and inhibited reproduction.

Family Trichosomoididae

This is an interesting family about which little known. *Trichosomoides crassicauda* is a parasite of the rat bladder, widely distributed throughout the world in wild rat populations and laboratory colonies. The adult female lives attached to the mucosa; the male is remarkable for living in the uterus of the female, 3–4 males occurring within a single female. The eggs hatch in the intestine, the controlling factors apparently being carbon dioxide, a low Eh and pH and the presence of bile (Chapman and Undeen, 1968) as is the case with many other eggs and cysts. The life is not known in detail; after hatching, released larvae probably pass to the lungs and are carried by the bloodstream to the bladder.

DIOCTOPHYMATIDA

Members of this order are large forms parasitic in the kidneys or occasionally the peritoneal cavity of mammals including man.

Chapter—8

Wuchereria (=Filaria) Bancrofti

The parasite gained importance after *Manson's* findings in 1878 that a mosquito acts as the transmitted agent for this parasite. *Manson* (1887) studied the life-cycle of the parasite and suggested reasons for the marked diurnal periodicity exhibited by filaria in the peripheral human circulation. *Vogel* (1928) and *Fain* (1951) gave accounts of the adult structure. The importance of this parasite to mankind is underlined by the fact four international meets were held in late 1960's to discuss the epidemiology, pathology and control of this worm.

SYSTEMATIC POSITION

Phylum	—	Aschelminthes
Class	—	Nematoda
Sub-class	—	Phasmida
Order	—	Filaroidra
Genus	—	*Filiaria*
Species	—	*bancıofti*

Name of Organisms and Disease Caused

Wuchereria bancrofti Silva Araujo, 1877 (syn. *W. pacifica* for the subperiodic form), and *Brugia malayi Bucklery 1958* (*syn. Wuchereria malayi*), cause Bancroftian filariasis, Brugian filariasis, periodic filariasis, subperiodic filariases, and elephantiasis.

Hosts and Location

Adults are found principally in lymphatic vessels of the extremities. Microfilariae (mf) are in the circulating blood.

Distribution and Importance

W. bancrofti is the ore widespread of the two worms; it is found in both the Eastern and Western Hemispheres, principally in the tropics, but with some extension into subtropical climates and small pockets in temperate areas. The areas of greatest prevalence are in Southest Asia, the islands of the South Pacific, Africa, the southern portion of the Arabian Peninsula, and India. Areas of lesser importance are in Japan and Korea, West Indies, South America, and areas fringing the Mediterranean Sea.

B. malayi has a limited distribution, principally in Southeast Asia, with more scattered distribution on the Indian subcontinent and on mainland China and Korea.

Stoll's (1947) estimate of filarial infections (*Wuchereria* and *Brugia*) in humans was 189 million persons. A more recent estimate by the World Health Organization (WHO) in 1974 was 250 million infections. In 30 years or so, little or no progress has been made in reducing the prevalence of infection.

Nearly everyone is familiar with the gross deformities that may be seen in long-standing infections with filaria in which the legs or genitalia are swollen and the skin has undergone irreversible changes. Such extreme changes occur only in persons living in endemic areas and who have been exposed and reexposed over many years. The prevalence of infection in endemic areas may exceed 50 per cent, as measured by microfilaremia, but the prevalence of elephantiasis is usually much lower. In some localities 10 to 20 per cent of the population may be affected; however, in Puerto Rico, 24 per cent of males were found to be infected, but the infection in nearly asymptomatic. Needless to say, severe deformity is a great burden to bear.

Morphology

The adult worms are found in the lymphatic vessels and lymph nodes of man only.

Adult worms are hair like, transparent nematodes often creamy white in colour. They are filiform and both ends and tapering. The posterior end terminates in a slightly rounded swelling. The male is about 40 mm, in length while thickness is about 0.1 mm. The posterior end of the male is curved ventrally and has two spicules of unequal length. Female measures 65-100 mm, in length and its thickness is about 0.2 to 0.3 mm. The tail end of the female is narrow unbent and abruptly pointed. Although females, liberate active embryos yet they are really *ovo-viviparous* i.e., lay eggs with well developed embryos. Males and females usually remain coiled and can only be separated with great difficulty. Females are more in number than males. The life span of the adult worm is about 4-5 years.

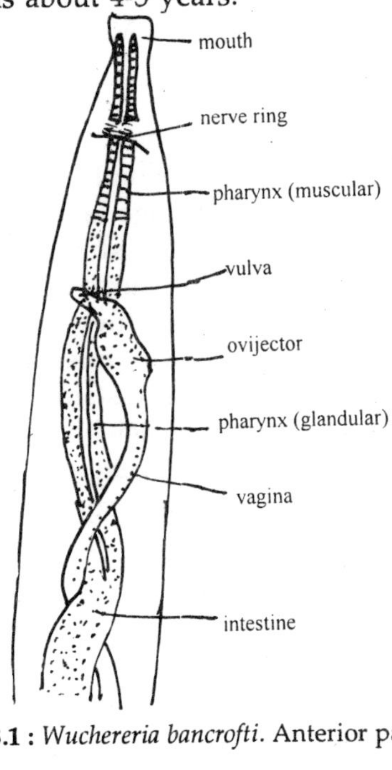

Fig. 8.1 : *Wuchereria bancrofti*. Anterior part.

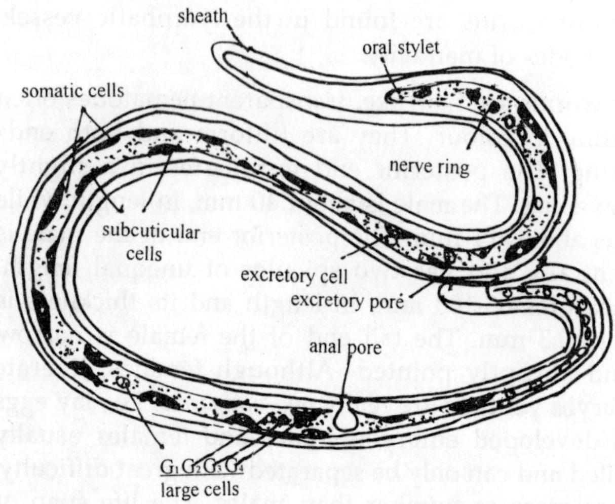

Fig. 8.2 : *Wuchereria.* Microfilaria.

Life-Cycle

Copulation takes place when individuals of both the sexes are present in the same lymph node. Female, releases numerous (about 1,000) first stage larval or jureinule called *microfilariae*. These larvae are surrounded by delicate membranes or sheaths and measures 225-300 μ long and 10 μ in diameter. When living they are colourless and transparent and may not be surrounded by sheaths. In order to identify them they are usually stained. The body will be found to contain a column of nuclei, broken in definite places. The peripheral ores are a nerve ring anteriorly, an excretory pore or 'V' spot, an excretory call somewhat further back, a few genital calls posteriorly and an and pore. The tail is devoid of nuclei.

The microfilairal appear in the peripheral circulation chiefly between 10 P.M and 4 A.M. There has been much speculation and experiment to determine the reason for this periodicity. According to one view, the larval are concentrated in internal organs such as lungs (Smyth, 1976) during day time, but they circulate in the blood at night for the transmission to the mosquitoes. Another view holds that embryos are born at a certain time each day and are then destroyed in the host within next 24 hours. According to *Manson-Bahr* and *Muggleton*, the nocturnal periodicity of microfilarial in peripheral blood of man is an adaptation to maintain a tryst with the night biting species of mosquito.

The further development of the microfilarial depend on their being sucked with blood by certain species of mosquitoes. In order to infect mosquitoes there must be about 15 or more microfilarial per drop of blood, a high concentration of 100 or more per drop of blood is fatal to mosquitoes.

Shortly after being disgested by a mosquito the microfilarial penetrate through the stomach wall and migrate to the breast muscles. Here the embryos lie length-wise between the muscle bibres. The size of body is reduced to half it original size length, but grows several times as thick. It now changes into a sausage-shaped creature.

The sausage shaped larva measures 1.5-2.0 mm. in length and 20-30 μ in diameter. As these changes take place the

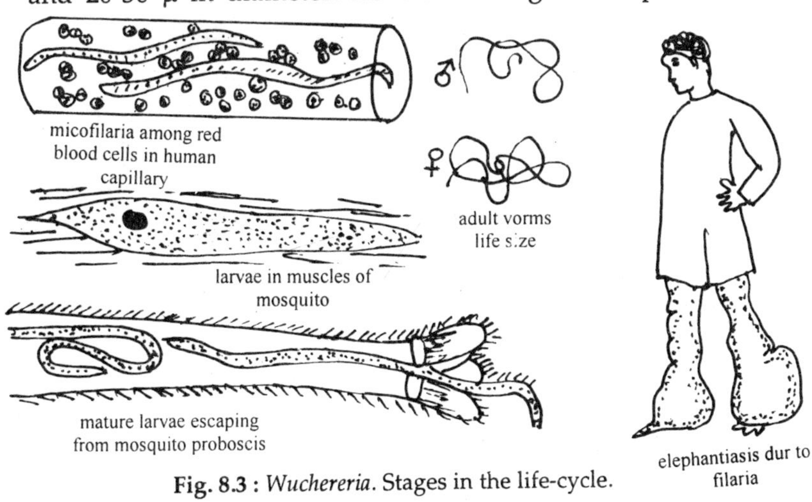

Fig. 8.3 : *Wuchereria*. Stages in the life-cycle.

larva mouths twice and become infective. The formation of the infective stage takes about 10-14 days and occurs best at 28°C and a relative humadity 90. Digestive system, body cavity and genital organs also become well developed. This is called third stage larva. This infective stage reaches into the head of mosquito, and finally into the proboscis on about 14th day. There are several larvae, coiled up in mosquito ready for infection in man. The complete development of microfilaria in the mosquito is from 10-20 days.

Infection and adult stage in man. When the infected mosquito bites a man, a third larvae are deposited, usually in pairs, on skin near the site of puncture. Later on the larvae enter through the puncture or penentrate through the skin. From here they reach the lymphatic vessels and settle down at same spot, i.e. inguinal, scrotal or abdominal lymphatics. The larva grows into adult. After a period of 5-18 months, larval become sexually mature and start new generation of microfilarial.

Diagnosis

Determining whether a person is infected with *W. bancrofti* or *B. malayi* depends on the following:

1. Clinical sings and symptoms.
2. Living in or having visited an endemic area.
3. Finding mf in the circulating blood.

Early in the infection there may be swelling of lymph nodes, skin rash, fever and urine cloudy with lymph (chyluria). Such signs occur in a variety of infections diseases and are not specific for filariasis.

Finding the mf in the circulating blood is good evidence when combined with clinical aspects and a high probability of having been exposed. Since there are a number of mf that may be found in humans, it is important to examine them with care and to identify them.

Failure to find mf does to mean that the person is not infected with the filarioid; in periodic filariasis, mf are not

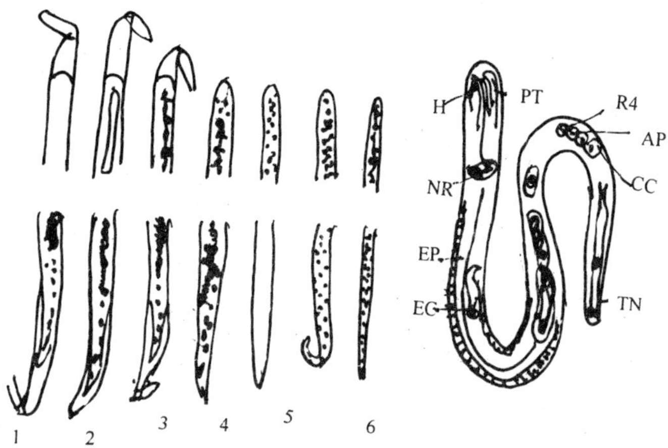

Fig. 8.4 : Morphology of microfilariae found in humans. (1) *Wuchereria bancrofti;* (2) *Brugia malayi;* (3) *Loa loa;* (4) *Onchocerca volvulus;* (5) *Dipetalonema perstans;* (6) *D. streptocerca;* (7) *Mansonella ozzardi;* AP=anal pore; CC=caudal channel; EC=excretory cell; EP=excretory pore; H=hook; NR=nerve ring; PT=pharyngeal thread; R2—R4=rectal cells; TN=terminal nuclei.

seen in the peripheral blood during the daytime. As indicated previously, the peak of microfilaremia takes place between 10 p.m. and 2 a.m., and it is important to take the blood sample during that peak time, or at least close to it. If a patient is hospitalized, taking a blood sample in the middle of the night is feasible, but appearing at a person's home at midnight with a syringe and needle at the ready is not likely to elicit enthusiastic cooperation. Epidemiologic studies as well are hampered by these factors. The DEC Provacative Test causes mf to appear in the blood 30 to 60 min after a patient has been given 100 mg of diethylcarbamazine (DEC); this eliminates the need for night sampling.

Microfilariae are found by using a compound microscope and one of the following techniques:

1. Stained blood film
2. Counting chamber
3. Membrane filtration

Table 1. Characteristics of Microfilariae

	W. bancrofti	*B. malayi*	*L. loa*	*O. volvulus*	*D. perstans*	*D. streptocerca*	*M. ozzardi*
Length	200-300 μm	177-260 μm	250-300 μm	285-368 μm or 150-287 μm	150-200 μm	180-240 μm	150-200 μm
Sheath	Present	Present	Present	Absent	Absent	Absent	Absent
Location	Blood	Blood	Blood	Skin	Blood	Skin	Blood and skin
Vector	Mosquitoes	Mosquitoes	*Clerysops*	*Simulium*	*Culicoides*	*Culicoides*	*Culicoides*
Periodicity	Nocturnal or subperiodic	Durnal	Present	None	None	None	None
Tail end	No nuclei Pointed up	2 widely spaced nuclei, blunt tip	Nuclei present, rounded tip	No nuclei, pointed tip	Nuclei present, rounded tip	Nuclei present, curved tip	No nuclei, pointed tip

The stained blood film has been the traditional method of finding mg; a thick film is best so that a relatively large volume of blood can be examined easily. The counting chamber method allows one to examine up to 200 mm^3 of blood but does not allow identification of the mf species. In the membrane filter method, blood drawn from a vein is passed through a filter which catches the mf. The membrane filter technique is generally considered to be the most sensitive.

Host-parasite Interactions

The damage that either *W. bancrofti* or *B. malayi* causes is dependent on the length and intensity of exposure to infective larvae and the response of the individual person. For example, military personnel were exposed to infection on some islands in the South Pacific during World War II for a period of a few months to 2 years. Many had signs and symptoms of infection, but the disease did not progress to the elephantoid stage.

For persons remaining in an endemic area, the disease progress through four stages:

1. Incubation
2. Symptomless, but patent
3. Acute
4. Chronic

As noted, the L_3s are about 1 mm long when they enter the definitive host. They move through the lymph system and are trapped in lymph nodes or in the tissue spaces. There is a reaction around the growing worms giving rise to inflammation (pain, swelling, and tenderness) and usually fever. Seemingly at about the time the worms mature, the symptoms subside and mf are seen in the blood (stage 2). This stage may persist for more than a year or may never progress further. The acute phase (stage 3) is caused both by significant blockage of lymph vessels and by the reaction of the host to the worms. There is swelling due to obstruction of lymph vessels, and a complex of symptoms and signs indicate that hypersensitivity has developed.

After some years, the swelling and obstruction of vessels leads to the irreversible changes in the extremities and genitalia that characterize the disease in the minds of most people. The sequence of events is basically blockage of the lymph vessels, causing retention of fluid distal to the blockage, a foreign body response, and an allergic reaction in the vicinity of the adult worms leading to additional swelling. The skin becomes progressively stretched and loses its normal structure.

The role of protective immunity in altering the course of filariasis is not at all clear. Antibodies are produced in the infected person; they can be detected by a variety of serological techniques, but there is not much evidence that the host is protected by the immune response. Also, the adult worms can live for 15 years or more, which indicates that they either are poorly antigenic or somehow escape the immune response.

Circadian periodicity is the typical pattern of microfilaremia in the definitive host. During the daytime, the microfilaremia is at such a low level that it is unlikely that any will be found in a sample of peripheral blood. During the evening, the mf begin to appear and peak usually between 10 p.m and 2 a.m.; then they begin to decline and disappear around dawn. Two questions that investigators have tried to answer are, where do they go and why do they cycle?

The answer to the first question is that the mf concentrate in the lungs during the daytime. The second question has two answers. There is a nocturnal periodicity, so that the mf appear in the peripheral blood at the time when there is greatest likelihood of their being ingested by mosquitoes that normally feed from twilight till dawn. How they know it is night time is another tissue, and most of the evidence points toward oxygen levels in the blood, which tend to decline when a person is asleep. Experiments have shown that the microfilaremia can be reduced at night if a person breathes pure oxygen.

In some geographic areas, the nocturnal periodicity of microfilaremia is not as striking; originally this condition was referred to as *nonperiodic filariasis* but is more correctly called *subperiodic filariasis,* because the day-night alteration is damped so that there is about a twofold increase at night rather than complete absence of periodicity. Subperiodic filariasis occurs with both *W. bancrofti* and *N. malayi* in localities where the principal vectors are day-biting mosquitoes. Subperiodic Bancroftian filariasis occurs in the Philippines, Fiji, Samoa, Tahiti, Tokelau, Wallis, and Ellice Islands. The localities of subperiodic Brugian filariasis are unclear, but seem to be based on the host in which it is principally found. The strains of *B. malayi* that cycle mostly in the human population are periodic, whereas those that cycle in feral animals with humans as an occasional host are subperiodic.

The question of why microfilarial periodicity occurs has been addressed but not answered. We can seen that adaptation to the time when mosquitoes tend to feed is advantageous. But why not be there all the time? It would not seem to cost the mf anything to circulate continuously, and they would be in an area to be ingested should a mosquito take a blood meal at lunchtime rather than at midnight. Evolutionary dogma stages that every characteristic of an organisms has adaptive advantage; it is not obvious what advantage is conferred by periodicity.

Treatment

Two different aspects of filariasis need to be considered: alleviating the effects of the infection for the individual and reducing the probability of transmission for the general well-being of the population.

The only drug recommended for treatment of either Bancroftian or Brugian filariasis is diethylcarbamazine (hetrazan, DEC). It is active against microfilariae, but there is still some doubt as to its efficacy against adults. It is taken by mouth three times a day for 10 to 30 days. In some instances there are side effects resulting from the death of the mf, and it is therefore difficult to ensure that people will continue to take the drug for the full course. At the end of the course of treatment, the microfilaremia will have been reduced completely, thereby eliminating the probability of transmission. The female worms will at least have been sterilized, and both males and females may have been killed. DEC was developed by Redginal Hewitt at the Lederle

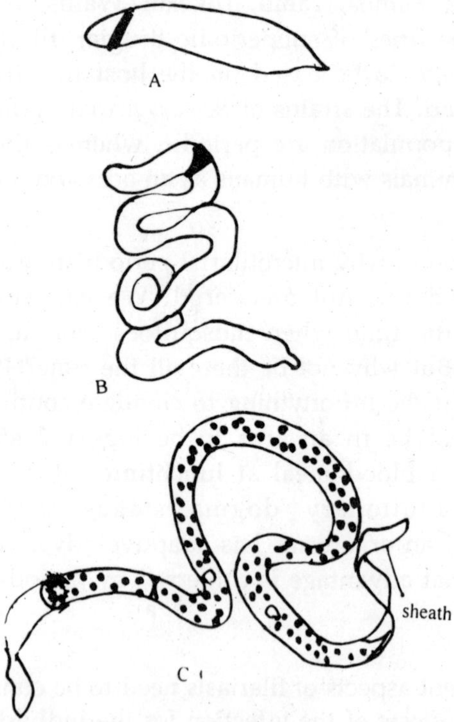

Fig. 8.5 : *Wuchereria bancrofti* larvae. A. sausage-share, stage, B. infective larva, C. microfilaria in blood.

Laboratories in about 1945; no better drug has appeared in the intervening 35 years or so.

The gross deformities seen in some cases of filariasis can be prevented by early drug treatment and by bandaging the legs so that fluid cannot accumulate. When the part has been badly affected, the changes are irreversible and surgery helps to reduce elephantoid deformities.

Surveys, ecological and population studies and the importance of mosquitoes as vectors of filariasis have been investigated in tropical Africa by Brengues (1973, 1975), Brengues *et al.* (1968), Diallo *et al.* (1978), Gratz (1973), Mattingly (1962), Nelson, Hsieh, and Furlong (1962), Subra (1975), Subra and Mouchet (1968), and White (1971). Also studies have been carried out on the genetics, susceptibility, etc. of mosquitoes by Kilama (1976), Magayuka and White (1972), Obiamiwe (1977a, 1977b), and Ogunba (1971a), among others.

Pathogenesis and Clinical Symptoms

The location of the adult worms in the lymph glands and connective tissues indicate that they would induce inflammatory and immune responses and, indeed, the pathogenesis is largely of this nature. Before patent microfilaraemia (*i.e.* before the microfilariae appear in the blood) the infection is symptomless. Then follows the acute stage marked by acute inflammation of various parts of the lymphatic system which become swollen and painful, accompanied by headaches, nausea, urticaria, elephantoid or filarial fever with temperatures reaching 40°C (104°F), and sweating. The inflammation of the lymphatic system manifests itself in lymphangitis, lymphadenitis (inflammation of the lymph ducts and lymph nodes respectively), lymphoedema, chyluria (lymph in urine making the urine milky in colour), lymph scrotum and hydrocoele, and lymph varies in which the lymph is blocked in the lymph ducts. If this condition persists over a long period (usually for years) the lymphoedema becomes chronic, especially in the lower, limphoedema becomes chronic, especially in the lower limbs, scrotum, vulva, breasts, and arms which become swollen and hard. There is fibrosis and hyperplasia of the tissues, and the skin thichens and sows folds and cracks in places, creating sites for secondary bacterial infection. Thus, the condition known as elephantiasis gradually develops.

It is now known that elephantiasis does not arise from blockage of the lymph ducts by adult worms, but is the result of allergic reactions to metabolic products of the worms over a long period. In fact, there are usually no microfilariae in the blood when elephantiasis sets in. In eastern Africa, particularly in Ethiopia, studies by Oomen, and Price have shown that in some areas where elephantiasis of the legs is endemic, *W. bancrofti* does not seem to occur, and so the condition cannot be due to bancroftian filariasis. Prince was able to show a link between the occurrence of elephantiasis and the type of alkaline volcanic soil common in Ethiopia. He therefore postulated that soluble silicates enter the body through wounds in the skin of the feet and legs, and crystallize in the lymph nodes, inducing inflammatory reaction which ultimately contributes to the onset of elephantiasis. Also, Castellani (1969) believes that elephantiasis is not directly caused by *W. bancrofti* since, even though filaricides may kill the worms, the recurring attacks of lymphangitis persists. He recognized many forms of elephantiasis and put forward the 'Castellani theory' which shows that *Elephantiasis nostras* (found in the temperature zone) and *Elephantiasis tropica* (found in the tropics and supposed to be caused by filariae) are essentially the same disease whose etiology is bacterial (streptococcal.) In the temperate zone, the bacterial infection is the primary cause while in the tropics it may be secondary cause, the filaria merely serving as an irritating agent paving the way for bacterial infection.

Diagnosis

This can be done by finding microfilariae is stained thick blood film. It is sometimes necessary to distinguish different species of microfilariae in the blood. The presence or absence of a sheath, the location of internal nuclei and primordia of organs particularly in the posterior end, the relative position of the nerve ring, etc., are reliable features. Gratama (1970) has produced a scheme for differentiating *Microfilaria bancrofti* and *Microfilaria volvulus* in the hydrocoele fluid in Liberia.

Haemagglutination and fluorescent antibody tests have proved successful for diagnosing chronic cases in which there is no microfilaraemio. Complement fixation and intradermal tests have not given reliable diagnosis.

Epidemiology

Some of the features of the epidemiology are summarized below:

1. The epidemiology of the disease depends on a complex of factors:
   (*a*) The distribution, local variations in density, susceptibility, longevity, and man-biting habits of the vectors.
   (*b*) The density, sleeping habits, and natural and acquired resistance of the human population.
   (*c*) Climatic factors such as high temperature and high relative humidity.

   So transmission should be seen as a dynamic system.
2. The vectors of urban and rural filariasis and their distribution and role in transmission in West and East Africa have been reviewed by de Meillon (1977) and treated above.
3. From all accounts, filariasis is increasing in prevalence and distribution in Africa, mainly because according to De Meillon:
   (*a*) The development of efficient waste disposal facilities has not kept pace with massive population migration to urban, suburban, and

industrial areas. The contaminated water in pit latrines, septic tanks, and drains therefore constitute good breeding habitats for *Cu.p. fatigans* and hence increase in prevalence of urban filariasis.

(*b*) The spread of rural filariasis, on the other hand, has been enhanced by the development of dams, reservoirs, etc., which form ideal all-reason habitats for the vector *A. gambiae.*

4. Filariasis is rare among children. The incidence increases with age, but it is observed that those who show microfilaraemia have been exposed to repeated bites of infected mosquitoes over a long period.
5. There are over 250 million cases of filariasis in the world, over half of which are in South-East Asia, particularly India. Because the prevalence of the disease is higher in East African than in West Africa, it has been proposed that the disease was introduced into Africa from South-East Asia via East Africa.
6. However, Wijers (1977b) suggests that bancroftian filariasis in Africa and India may be caused by two different strains or the worm, because of difference in the parts of body affected. In East Africa, hydrocoele is a common feature of the disease, over 50 per cent of infected males being affected. In India, on the other hand, elephantiasis of the legs predominates over hydrocoele. In West Africa, though high microfilarial rates (40 to 50 per cent) may be encountered, there are only a few clinical manifestations of the disease, suggesting the possibility of a third strain.
7. The absence of microfilariae in chronic cases suggests the development of immunity. Thus reliance on blood smears for microfilarial rates

alone may give a misleading picture of the epidemiology.

8. Also, misidentification of microfilariae can give a misleading idea of the epidemiology. For example, Nelson, Hsieh, and Furlong (1962) found microfilariae of six different species of animal parasites in the mosquito *Aedes pembaensis* which was previously thought to be the vector of *W. bancrofti* in Kenya. Similarly important is the accurate identification of the vectors, but there is no international reference centre for mosquitoes.
9. Accounts of the epidemiology of bancroftian filariasis and transmission by mosquitoes have been given in various publications by Brengues and Subra and their co-workers and Diallo and colleagues in West Africa and Wijers in East Africa (already cited), as well as by Kuhlow and colleagues in Liberia and McConnell, Ashafa, and Dennis (1976) in Ethiopia.

Prevention and Control

Control follows the general programme for mosquito control treated in greater detail later under malaria control. The prospect is, however, gloomy because of resistance to insecticides. Larviciding of pit latrines and drains may give good results, but better still is the provision of adequate sewage disposal facilities which unfortunately is beyond the means of countries in endemic areas, mainly because of the prohibitive costs. Control of rural filariasis is much more difficult. Mass chemotherapy of the population is an attractive proposition but fraught with all kinds of problems, seemingly insurmountable. In short, 'up to the present date, effective measures of control of bancroftian filariasis have not been practical in Africa'.

Prevention of mosquito bites by the use of mosquito nets and insect repellants can also help. This could reduce transmission to such low levels that even though microfilariae might still be present, the more sever form of hydrocoele and elephantiasis will hopefully become a rarity.

GENERAL BIOLOGY

Onchocerca volvulus is a filarial worm which infects over 40 million people in the world, some 30 million of them living in the greater part of tropical Africa. Other endemic areas are the Yemen, Guatemala, Mexico, Venezuela, and Colombia. The parasite causes the disease onchocerciasis, one of the most disabling and debilitating parasitic diseases known to man. The characteristic presentation of the disease is dermatitis which not only leads to disfigurement and premature ageing in appearance but the constant pruritis results in incessant scratching by individuals and sometimes whole communities—an activity which prevents steep and detracts from productivity. If that were all, the condition though causing great unhappiness could be endured and indeed has been accepted by some of the victims as part of the progress of ageing. However, the parasite causes 'river-blindness' responsible for the highest blindness rates in the world. In some communities in endemic areas in West Africa, blindness rates due to onchocerciasis is over 10 per cent compared to 0.5 to 1.0 per cent in non-endemic parts of the same geographical area.

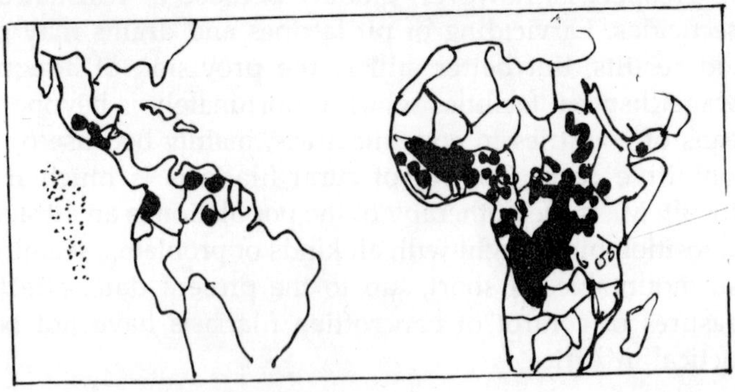

Fig. 8.6 : Map showing the distribution of *Onchocerca volvulus* in Africa and South America.

The disease is of great economic importance in tropical Africa because not only does it incapacitate a large segment of the adult working population and prevent them from reaching their maximum productive capacity but it has prompted whole communities to abandon fertile lands near river valleys in favour of less productive areas in order to be free from the ravages of the disease. Indeed, onchocerciasis has been grouped with trypanosomiasis and malnutrition as the factors which have frustrated the economic development of many of the already impoverished savanna regions of Africa.

It is clear therefore that though onchocerciasis is an economically important disease, it is not a killing disease. In the words of Nelson (1970):

In spite of its obvious importance, onchocerciasis has been neglected and until comparatively recently very little has been done towards its control. The reasons for this are as follows:

1. The victims of the disease are usually illiterate poverty-stricken peasants in remote rural areas who hardly 'make a hue and cry' about their plight and, even if they did, they do not wield sufficient influence to attract the attention of government and public health officials to do something about it.
2. Until comparatively recently, only a handful of scientists, physicians, and ophthalmologists have shown any interest in, or carried out any research on, the natural history, management, and control of the disease. Thus knowledge about these aspects has been limited.
3. Although about 2000 infected persons are reported to be living as immigrants in London, the disease is usually restricted to people who are exposed to repeated bites of the vectors in endemic foci. This is because the disease becomes manifest when the parasites accumulate in the body over long periods of time, measured in years. As a result, the disease

is hardly encountered outside the endemic areas and so has attracted very little attention in the outside world.

4. Studies on the pathology and pathogenesis of the disease have been hampered by the lack of suitable animal models for experimental work.
5. Effective drugs have not been developed because the pharmaceutical companies, prompted mostly by the profit motive, have not found the commercial prospects of developing drugs for the use of 20 million poor peasants an attractive proposition.

Morphology

The adult worms show most of the characteristics of the order Filariata. The males measure 19 to 42 cm long and 130 to 210 μm wide, while the females measure, 33 to 50 cm long and 270 to 400 μm wide.

Adult worms are knotted together in the subcutaneous tissues where tissue reaction results in the formation of nodules round them. The microfilariae are unsheathed and remain under the skin.

Many biological variants or strains have been recognized, although they are indistinguishable morphologically. Thus on epidemiological and pathological ground differences have been found between the African and American forms of *O. volvulus*. In West Africa, Duke and his colleagues have established the existence of two distinct strains, the 'forest' and 'savanna' strains differing in epidemiology, clinical features, vector infectivity, etc., while more recently, Braun-Munzinger and Southgate (1977) have shown four sympatric types of *O. volvulus* in Togo by showing differences in the enzyme acid phosphatase profiles of the microfilariae.

Life Cycle

The nodules containing the adults are usually located above the waist in Central America and below the waist in

Africa, probably an adaptation to the differences in the biting habits of the respective insect vectors. The vectors are various species of the blackfly *Simulium*. Their mouth parts are modified for feeding on tissue juices of animals including man, from just under the epidermis. When they bite an infected man, therefore, they ingest microfilariae along with the tissue juices.

The development in the fly follows the same pattern as in other filariae. Nelson and Pester (1962) and Duke (1968 b) have described the developmental stages in the fly. The first stage larva migrates to the thoracic musculature where it moults to form the sausage shape. This moults into the third stage larva which is filariform, then positions itself around the labium, from where it gains entry into a new host when the insect feeds. There is evidence of transplacental transmission of microfilariae so that the foetus is infected *in utero*. The route taken by the worms to their definitive location in man is still not known. Similarly, it is not known why their preferred sites are over bony prominences.

The life span of the adult worm is estimated at 16 years while the microfilariae can survive for 30 months in the skin.

ONCHOCERCIASIS—PATHOGENESIS AND PATHOLOGY

The pathogenesis of onchocerciasis arises from two factors: effects due to the adult and those due to the microfilariae. The adult effects are minimal in early stages or light infections, the adult worms lying freely in the subcutaneous tissue. In order infections, the worms induce a tissue reaction of the host to produce the characteristic subcutaneous nodules (onchocercomas) around them. These palpable nodules are usually sited over bony prominences and in Africans are distributed mainly below the waist, especially in the pelvic region but also along the spine, chest, and knees. The nodules vary in size and numbers (up to a hundred may be found in a single patient), but their effect is benign.

The effects of the microfilariae are far more serious and may be due to allergic reactions of the host to dead microfilariae. Nelson (1970) and Buck (1974) have given accounts of the course of the disease summarized as follows:

1. There is first the production of papular rash (craw-craw), a condition known as onchocercal dermatitis or onchodermatitis characterized by intense itching and scratching, usually resulting in secondary bacterial infection.
2. This is followed by pigmentation changes consisting of hyperpigmented spots with depigmented centres.
3. Thickening of the skin due to intradermal oedema which when localized produces the 'peau d'orange' effect.
4. This is followed by lichenification of the skin such that a mosaic pattern results (pachydermia), giving the patient 'lizard' or 'crocodile' skin.
5. The last stage is characterized by loss of elasticity and atrophy of the skin (presbydermia), giving the patient a prematurely aged appearance. At this stage, depigmentation becomes more extensive and when the shins of the legs are involved, the 'leopard skin' effect is produced a condition which may be and is, often confused with leprosy.

Most of the symptoms in onchocerciasis are similar to those produced in avitaminosis A. It is possible that the parasite competes for vitamin A in the skin or interferes with its metabolism. On the other hand, it may be that people suffering from vitamin A deficiency are more susceptible to heavy infection. Recently, Bonucci, Brinkmann, and Onori (1979) have studied the histopathological changes in the skin on onchocerciasis patients to Togo.

Other effects of microfilariae include the following:

1. When the loss of elasticity occurs in the femoral region, it results in the condition known as 'hanging

groin' in which there is sagging of the groin into pendulous sacs. These sacs may contain lymph nodes, particularly the femoral and inguinal lymphnodes, but testes and scrotum are not involved.

2. According to Nelson (1970) onchocerciasis is one of the most important causes of hernia, especially femoral hernia in Africa.
3. Onchocerciasis has been associated with elephantiasis in some part of Africa. Oomen's (1967) work in Ethiopia showed that scrotal elephantiasis due to onchocerciasis was prevalent, but this could be distinguished from the elephantiasis of the scrotum due to *W. bancrofti* because, in the former, there is usually no accompanying elephantiasis of the lower extremities. So far there is no conclusive evidence that onchocerciasis is a cause of elephantiasis of the lower extremities.
4. Dwarfism has been associated with severe infection with *O. volvulus* in Uganda. The pathogenesis of this condition is not fully understood, but it is believed to be due to pituitary deficiency resulting from damage of the pituitary gland by microfilariae. The Nakalanga dwarfs are not true pygmies because they are offsprings of normal parents.

Ocular Lesions

The most serious manifestations of onchocerciasis are the eye lesions which result in the much-dreaded river blindness. Total blindness or even impaired vision takes many years to develop and usually become evident in people in their forties. It also depends on the density of microfilariae in the skin, developing in patients in whom the microfilarial density in the lower part of the body is as high as 100 per

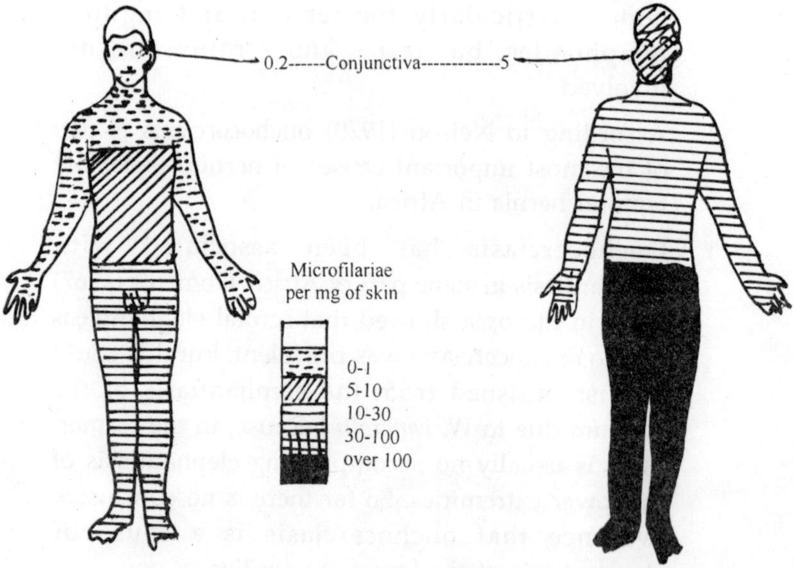

Fig. 8.7 : Diagram to show the distribution and density of microfilariae of *Onchocerca volvulus* in different parts of the body in relation to ocular involvement: A. no eye lesions, B. eye lesions present.

mg of skin. Kershaw, Duke, and Budden (1954) have prepared a chart of the distribution and density of microfilariae, as it relates to eye lesions in West African onchocerciasis. Accounts of the pathology of ocular onchocerciasis have been given by Anderson and Fuglsang (1977), Buck (1974), Choyee (1966), Duke (1972b), Nelson (1970), Rodger (1962), and Thylefors (1978).

Briefly, ocular involvement arises from invasion of the eye by microfilariae from the skin through the conjunctive, and the immune responses to dead microfilariae as well as to their toxic metabolites. Three syndromes are readily recognizable:

1. Large numbers of microfilariae in the conjunctiva cause a chronic brick-red conjunctivitis.
2. The most characteristic lesions are produced when large numbers of microfilariae invade the cornea. When they die, they cause inflammation, followed by fibrous tissue invasion which produces an opacity in the lower half of the cornea which impairs vision. The gradual spread of this opacity over the entire cornea leads to complete blindness. During this period, microfilariae can be seen in the anterior chamber of the eye with the aid of a silt lamp. When they die in the iris and ciliary body, they produce a chronic iridocyclitis.
3. Sometimes, microfilariae in the posterior segment of the eye may cause choroid-retinitis and atrophy of the optic nerve which results in reduced visual fields and ultimately blindness, although onchocerciasis may not be the only cause of this condition.

It is estimated that half a million people are presently afflicted with river blindness, half of that number being completely blind. Another half a million stand the risk of a similar fate in years to come. Other statistics show that in some communities in endemic areas of the savanna in Africa, while 30 per cent of the population may have impaired vision, 10 per cent of them will be completely blind, including 25 to 50 per cent of the adult male population who are the most productive members of the community. Another interesting feature of river blindness is that the incidence in the drier savanna regions of Africa is more than ten times that in the rain forest region, even though transmission may be just as intensive, if not more so. Anderson *et al.* (1974a) showed that in persons 40 years and above in Cameroun blindness due to onchocerciasis reached 4.2 per cent in the rain forest and 14.4 per cent in the savanna. In support of this, Ayanru

(1974, 1977, 1978) who has studied blindness extensively in the Bendel State of Nigeria concluded that: onchocerciasis (and trachoma) which cause much ocular morbidity and blindness in the savanna north have little place in the rain forest areas of southern Nigeria; neither has it is the derived savanna area of Ibarapa and Lagos in south-western Nigeria.

Rolland, Thylefors, and Pairault (1978) have studied the progress of ocular onchocerciasis in a village community in Upper Volta (where the disease is hyperendemic) over a nine-year period. Of those previously diagnosed to have eye lesions, 50 per cent were found to be blind nine years later and 40 per cent of those over ten years had reduced visual acuity. Above all, prognosis was poor because in the majority of the subjects, microfilariae were easily detected in the anterior chamber of the eye.

ONCHOCERCIASIS—DIAGNOSIS AND TREATMENT

Diagnosis is done by examining skin-snip biopsies under the microscope for microfilariae. A thin silver of skin of about 3 mm in diameter is sliced off with a razor, placed in water or physiological saline, and examined for microfilariae emerging. If the snip is weighed, teased, and the microfilariae counted, a quantitative estimate of their density, expressed as microfilariae per milligram can be obtained. There is danger of contamination with blood if the snip is too deep which could lead to confusion with microfilariae of other filarids like *Loa loa* and *Dipetalonema perstans*. The following precautions must be observed when using this technique:

1. It may be necessary to examine several skin-snips before obtaining a diagnosis in light infections.
2. Microfilariae may be so deeply situated in old infections that the skin-snip may not reach them.
3. If only a single snip is possible, it should be taken from the buttocks.

There have been reports of microfilarial periodicity of *O. volvulus* in the skin in Africa. It is likely, therefore, that

A

B

C

D

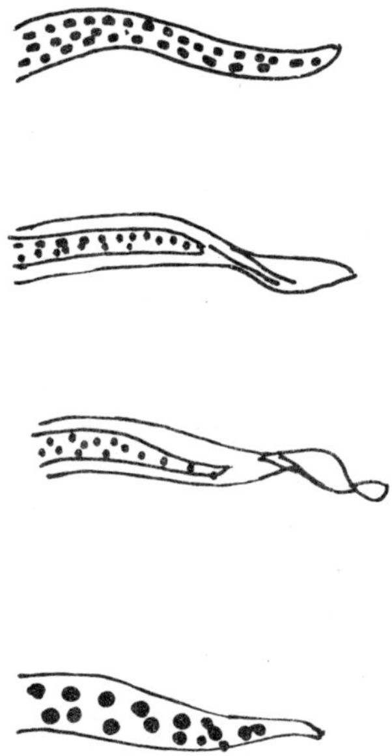

Fig. 8.8 : Posterior end of microfilariae of various filariae: A. *Dipetalonema perstans,* B. *Wuchereria bancrofti,* C. *Loa loa,* D. *Onchocerca volvulus.*

the microfilarial densities obtained will depend on the time of day, with obvious implications in epidemiological surveys. However, the periodicity appears to vary in different parts of Africa, depending on the peak biting activity of the local vectors. Thus, for example, Lartique (1967) found higher densities at 4.00 p.m than at 10.00 a.m. in West Africa, while Thomas, Anderson, and MacRae (1973a) observed the opposite in northern Cameroun. However, Duke *et al.* (1967a) found that these daily fluctuations are too small to make any appreciable difference in epidemiological studies.

Immunodiagnostic tests using antigen from various filarial worms have been positive for onchocerciasis. Unfortunately, however, they are non-specific and give cross-reactions with other filarial infections. The Mazzotti test is highly sensitive even in cases of light infection. It involves the administration of a dose of 50 mg of diethycarbamazine (DEC). The test is positive if there is an itching popular reaction in the skin within 1 to 24 hours. Oomen (1967) has found the Mazzotti test to be unreliable in Ethiopia.

Finally, diagnosis may be obtained through urine examination because microfilaruria has been found to be widespread in the onchocerciasis belt of tropical Africa.

Treatment

Surgical and chemotherapeutic methods are available for treatment of onchocerciasis. Nodulectomy (surgical removal of palpable nodules, especially those located in the head region) which has been effective in Central America has not produced encouraging results in Africa. The location of the nodules below the waist, and the fact that in Africa they contain mostly dead or dying worms, may account for this. Drugs are available for killing either the microfilariae or the adults, but they produce such unpleasant and severe side-effects that they are unsuitable for mass treatment. Indeed, according to mass treatment to Duke (1976), 'many persons with asymptomatic infections are best left untreated.'

Diethycarbamazine (DEC or Hetrazan) is a highly effective microfilaricide, but the allergic reaction which follows the death of the microfilariae (marked by severe pruritis and dermatitis) may be so violent as to result in extreme prostration and even death due to anaphylactic shock. Its use for mass treatment campaign is therefore not recommended. Duke (1968d) has studied the use and effects of this drug.

Suramin, on the other hand, is an effective macrofilaricide, killing only the adult worms. However, its incidental effects on microfilariae produce such severe reactions that it is extremely dangerous for mass treatment. However, good results can be obtained under medical supervision, the microfilarial density being first reduced by treatment with DEC before the administration of Suramin.

The application and effects of these drugs have been studied by Duke (1968d, 1972b).

Melarsen W (Mel W) also kills the adult worms of sterilizes the females, but it is very toxic. Duke and his colleagues have tried many other drugs but they are either not effective or are too toxic. Kale (1978) has shown that six drugs in common use for treatment of parasitic infections in man produced no evidence of activity against microfilariae or adults of *O. volvulus*. There are thus at present no practical chemotherapeutic agents for onchocerciasis and no vaccine has been developed. Progress in the reach for suitable microfilaricides has been hampered by the lack of alternative small mammals susceptible to *O. volvulus* that can be used as laboratory model for screening drugs. The prospects of findings a drug which will combine macrofilaricidal activity with lack of serious toxicity therefore remain bleak.

ONCHOCERCIASIS—EPIDEMIOLOGY

The epidemiology of onchocerciasis depends on the following:

1. The availability of the right species of *Simulium* vectors, the conditions for their development, their dispersion, and behavior.
2. The factors affecting the development of microfilariae in the files, their effects on the flies, and their transmission pattern.
3. The characteristics of the human population exposed to infection, the size of the population, their behaviour, occupation, and movement pattern.

A review of the epidemiology of onchocerciasis has been given by WHO (1976).

The Role of Insect Vectors in the Epidemiology of Onchocerciasis

Considerable knowledge about the ecology and biology of simuliids has been accumulated in tropical Africa. But Duke, who has contributed most to our knowledge in this field, lists three main hindrances to the practical study of *Simulium* biology:

1. The failure to establish self-perpetuating laboratory colonies of *Simulium;* the flies do not mate in captivity.
2. The need to develop new sampling methods for *Simulium* vectors.
3. The need to develop standard sampling methods for the developmental stages to permit accurate qualitative assessments to be made of the considerable natural fluctuations in their populations.

Nevertheless, the following features of the entomology are of epidemiological significance:

1. *Simulium damnosum* is the main vector of *O. volvulus* in tropical Africa. This is a complex of species, some of which are wholly zoophilic, and thus of no importance to man. They breed on vegetation and rocks in fast-flowing rivers and streams in the various vegetational zones of Africa. *Onchocerca volvulus* is also transmitted in eastern and central Africa by the *S. neavei* complex. These bread in small streams in highlands and maintain an obligatory phoretic association with specifics of the freshwater crab *Potamonautes* on which the larvae and pupae are found. For details of this association see van Someren and McMohon (1950) and Williams (1968). The taxonomy, biology, and ecology of *Simulium* spp. have been extensively studied by Dang and Peterson (1980), Grams (1978), LeBerre (1966), and Lewis and his colleagues (1948 to 1968).
2. There are different strains of the species of *Simulium* vectors. These strains tend to be adapted to the local strain of *O. volvulus* which they transmit. Thus members of the *S. damnosum* and *S. neavei* complexes tend to be low biters, which is an adaptation to transmitting the African strains of

O. volvulus which are located below the waist. On the other hand. *S. ochraceum* which transmit the American strain bite above the waist. It is also observed that microfilariae of strains of *O. volvulus* develop best in the appropriate strain of the simuliid. Thus distinct parasite—vector complexes are developed which have a geographical pattern of distribution. Thus, for example, there are two main parasite—vector complexes in West Africa—one in the Guinea savanna and the other in the Sudan savanna zone. Duke and his colleagues have studied this problem extensively.

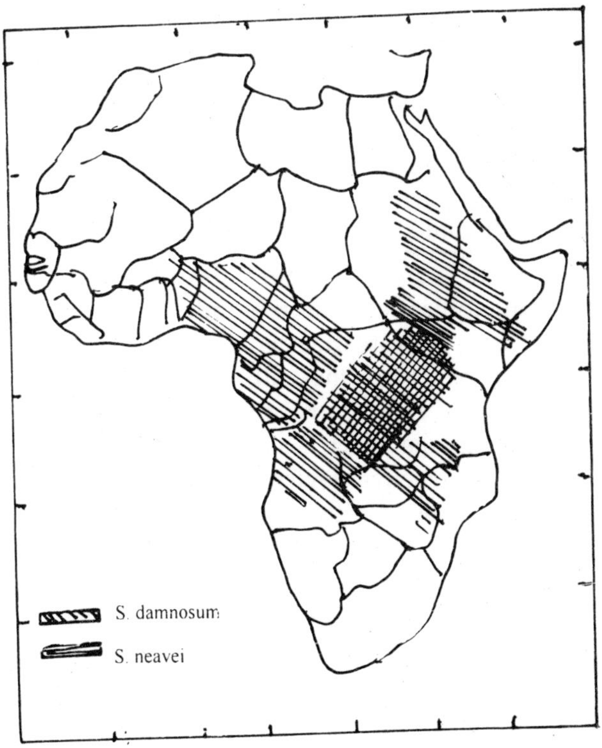

Fig. 8.9 : Map of Africa showing the distribution of *Simulium damnosum* and *S. neavei*.

3. It has been shown that a particular strain of the mosquito *Aedes aegypti* will support the development of *O. volvulus* experimentally.

4. The flies have to maintain a close association with fast-flowing water suitable for the development of their eggs, larvae, and pupae. This limits their distribution to the breeding sites and the area within the flight range the adult flies. Thompson (1976), reviewing records of flight range of flies in the savanna region of West Africa, showed ranges of 12.9 to 96 km, although the effective range is in the lower half of the scale.

5. The physico-chemical factors of the water will obviously play a part in the distribution and development of the insect. This has been studied in Ivory Coast by Quillevere *et al.* (1976, 1977a) who showed that the pH is the only important factor affecting the larval distribution of the different species of the *S. damnosum* complex.

6. Flies usually ingest enormous numbers of microfilariae when they bite infected human host. These microfilariae would cause high mortality of the flies were they all to develop to the infective stage. However, the chitinous peritrophic membrane which is secreted around the blood meal is a protective device which traps the microfilariae and prevents them from reaching the thoracic muscles. Thus only about 50 per cent of the ingested microfilariae survive and migrate within the first 24 to 48 hours; the rest die and disintegrate in the mid-gut. Up to 20 to 30 larvae may develop to the infective stage in wild *S. damnosum*.

7. Although the presence of the larvae in the thoracic flight musculature may reduce the flight range, there is no conclusive evidence that this affects the epidemiology appreciably.

8. A number of factors influence the efficiency of transmission. According to Duke (1972a), these include:
   (*a*) The absolute numbers of the vectors available,
   (*b*) Their longevity,
   (*c*) Their degree of anthropophily, and
   (*d*) The climatic and environmental conditions prevailing.

The index for measuring the absolute amount of transmission in unit time is expressed as annual transmission potential (ATP) which is 'the number of infective larvae to which a man would be exposed under optimal biting conditions in one year'. This index is useful for comparing transmission in different environments and for evaluating transmission before and after vector control operations in the same environment.

Quillevere *et al.* (1977b, 1978) have studied the ecology and vector potential of members of the *S. damnosum* complex in the savanna and forest zones of Ivory Coast.

Microfilariae in Epidemiology

1. Development to the infective stage in the fly takes 6 to 12 days depending on the ambient temperature. The development is also synchronized with the gonotrophic cycle in the insect, which is also dependent n temperature such that at the time the fly is ready for its third blood meal, the larvae have reached the infective stage and migrated to the proboscis from which they are transmitted at this and subsequent feeds.
2. Because members of both *S. damnosum* and *S. neavei* complexes have been found to be catholic in their taste rather than being strictly anthropophilic, it is not surprising that many of the infective larvae found in them are not *O. volvulus* larvae.

Misidentification of larvae will clearly affect the ATP index. It is not easy to distinguish *O. volvulus* from other filarial larvae in simuliids. Some of the distinguishing feature have been illustrated by Nelson (1970).

The Role of Man in the Epidemiology of Onchocerciasis

As in the case of insect vectors, knowledge about the epidemiology of onchocerciasis in man is fragmentary, inadequate, and in most cases deficient. According to Buck (1974):

1. There is paucity well-planned longitudinal studies in the general population in endemic areas.
2. While there is a wealth of cross-sectional data on geographical distribution of the disease and its pathogenesis, the procedures for obtaining these data have not been standardized and the populations studied not adequately defined in terms of age, sex, and other parameters. These deficiencies have been a source of frustration in interpreting and comparing these data.

However, some of the important features of the epidemiology as it relates to man are summarized as follows:

1. Man is the only known definitive host, no reservoir hosts having been found naturally infected. So far the chimpanzee is the only suitable experimental animal.
2. *Onchocerca volvulus* is remarkable in that the adults induce only a controlled tissue reaction in the host such that the fibrous nodules formed do not isolate and destroy the worms as in other defence mechanisms, but rather serve to protect them from physical damage.
3. Microfilarial density is greatest in the lower part of the body in Africa and it is known that the serious complications of the disease, and in

particular river blindness, develop only when exposure to infection has been prolonged over many years, resulting in high concentration of microfilarial spreading throughout the body including the eyes. Thylefors, Philippon, and Prost (1978) have studied the relationship between the quantity of transmission and the corresponding pattern of the disease in the Sudan savanna area of West Africa. They found that intensity of infection increased with ATP, and this was associated with an increase in incidence of blindness and ocular lesions. Consequently, short-term visitors and tourists to endemic areas are unlikely to suffer from the more serious aspects of the disease. Furthermore, the disease is progressive even in the absence of continued of continued transmission.

4. Transmission patterns are influenced by a complex of factors including vegetation, and occupation, and size of human population. Thus, DeSole and Kloos (1976) found three patterns of transmission in Ethiopia:

   (*a*) The forest area, although providing ideal breeding conditions for *Simulium*, had a low transmission rate because of sparse human population.

   (*b*) The subsistence farming area offering less favourable conditions for the vector nevertheless had a higher prevalence than the forest because of the larger human population.

   (*c*) The coffee farming area with its heavy rainfall, numerous rivers, and adequate shade (the most ideal conditions for the vectors) and dense human population provided the most favourable conditions for transmission to occur, and as expected a high prevalence of the disease was observed.

5. In addition to differences in vector infectivity between the savanna and rain forest strains of *O. volvulus* in West Africa, there are also marked differences in intensity of infection and clinical manifestations in man. Anderson *et al.* (1947b) who have studied this problem extensively, compare seventeen parameters observed in Sudan savanna and rain forest areas of Cameroun, and suggest that the observed differences may be due to variations in host response resulting from differences in the intensity and patterns of transmission, as well as changes in immunological response due to concomitant infections.
6. The behavior of man in response to the pathological effects of *O. volvulus* plays a significant role in the epidemiology of the disease. In the first place, the need to be rid of the disease may prompt man either to initiate control measures, or retreat from, or desert the transmission foci in the river valleys. Both of these options break the transmission cycle and are of no ecological advantage of survival value to the parasite. Duke (1972a) has given a succinct account of Hunter's (1966) hypothesis of cyclical growth and desertion of human settlement associated with onchocerciasis in riverine areas in the savanna of West Africa, summarized briefly as follows:
   (*a*) The fertile land in the river valley attracts human settlement as well as serving as ideal breeding sites for *S. damnosum*.
   (*b*) The local *S. damnosum* soon becomes highly anthropophilic and if *O. volvulus* is imported into the valley by the new settlers, it is easy to see how this is transmitted and spread in the population.

(*c*) It may take up to 20 to 30 years before the build-up in microfilarial density is sufficient to cause serious symptoms of the disease and 40 to 60 years before blindness becomes a problem in the community.

(*d*) After another 10 to 20 years, the effects become so serious as to affect the community so adversely as to compel them to abandon the valley.

(*e*) Future generations may avoid the valley, but eventually the search for fertile and productive land will lead people to the valley once more, especially as there may no longer be any evidence to indicate that the land had once been settled and abandoned. So the cycles starts all over again.

7. Water development projects like construction of dams for water supply, irrigation, and hydroelectricity may ostensibly alter the ecology by changing a flowing river into a lake, thereby rendering it unsuitable as breeding habitats for simuliids. Unfortunately such projects also create new breeding sites in the fast flowing water of the spillways, further enhancing the rate of transmission.

8. Ignorance about the natural history of the disease may be a factor in the epidemiology of onchocerciasis. Imperato and Sow (1971) report that in the Senegal River Basin in Mali, the people do not associate the signs and symptoms with the disease processes, considering them as aspects of normal anatomy or part of the ageing process, or dismissing them with some superstitious explanations. Though they realize the relationship between blindness and nearness to water courses, they are ignorant of the role of simuliids in

transmission of the diseases. On the other hand, people in endemic areas of Kenya were well aware of the relationship between skin disease and eye lesions, and knew that *S. neavei* in riverine areas were responsible long before the role of simuliids was officially discovered by Blacklock in 1926.

9. The pattern of village abandonment observed in the savanna belt of West Africa correlates strongly with the geographic pattern of onchocerciasis and blindness. In a detailed study of the problem in the Middle Hawal Valley in Nigeria, Bradley (1976) suggests that onchocerciasis is only a part of a complex of desertion, though it is the most important factor. This has far-reaching effects on the agricultural production and economy of the people by forcing them to desert rich fertile and productive lands in favour of upland, watershed areas, consequently resulting in overpopulation of these less productive areas, with the attendant over-cropping.

ONCHOCERCIASIS—PREVENTION AND CONTROL

Control of any parasitic disease involves measure directed at the parasite, the vector, and man. In the case of onchocerciasis, measures against the parasite are limited to nodulectomy and chemotherapy, both of which have been discussed above. The prospects of mass control by chemotherapy are not very bright because of the severe toxic side-effects of the drugs currently available. If man could avoid contact with the flies, the transmission cycle could be broken—an effective method of control. Traditionally, this has been achieved in Africa by literally running away from the problem. This has entailed the desertion of fertile but infested land in the river valleys with the concomittant economic and social problems already discussed hardly a solution to be recommended. The association of onchocerciasis with rich fertile lands makes it imperative that control should not only achieve freedom from the scourage but also bring about the transformation of these areas into safe and inhabitable settlements in which economic activity can thrive.

To achieve this, the only options open so far have been to attack the blackflies. Several approaches have been tried and different programmes mounted in various parts of Africa with various degrees of success. Brown (1962), Le Barre *et al.* (1978), McMohan (1967), Raybould and White (1979), and Waddy (1969) have reviewed the control of *Simulium* vectors of onchocerciasis in Africa.

Selective bush clearing was found to be effective against *S. neavei* in Kenya but was not applied on a large scale and was soon abandoned because of the high cost. Also, modification of the habitat by converting a flowing river into a lake has been tried. The dangers associated with the spillways can be eliminated by controlling the water flow and applying larvicides. However, the use of insecticides, particularly directed at the larvae, has proved to be the most promising approach. DDT has been found to be the most effective larvicide and has been used for control programmes in different parts of Africa with some success, *e.g.* Jinja on the Nile in Uganda, Mayo Kebi in Chad, Kinshasa around the River Cango in Zaire, Abuja and Kainji in Nigeria, the Volta River and northern Ghana, and the whole of Kenya. A few of these attempts will be reviewed here:

1. The Abuja programme showed that larviciding 160 km of river with DDT reduced *S. damnosum* density by 90 per cent of its pre-control level in an area of over 3000 km^2, but showed no obvious decrease in the prevalence of onchocerciasis over a ten-year period (1956 to 1966) attributable to the control measures, although the severity of infection had been reduced. It appears that in a small area within a larger endemic area, considerable reduction in fly density can be achieved but the low density attained can still maintain the transmission level. Imevbore *et al.* (1978) have made a survey of blackfly larvae in the area.

2. The Kainji control programme initiated after the dam was built also showed a reduction of 90 per cent vector population. However, surveys by skin-snip biopsies in an area which was hyperendemic before the dam was built showed a prevalence rate of 93 per cent and blindness rates of up to 9.0 per cent although Mellink and Gillieron (1977) observed that the reduction in prevalence was less than they had been led to anticipated.
3. The most successful of all is the Kenya project in which *S. neavei* was successfully eradicated from Kenya by McMahon and his colleagues probably the only example in the world in which a disease and its vector have been completely eradicated in an entire country. The project, extending over a ten-year period, covered rivers and streams in an area of over 8500 km^2, with a population of over a quarter of a million, and involved meticulous surveys, application of DDT as larvicides, and assessment of its effects. Surveys by Nelson and Grounds (1958) and Roberts *et al.* (1967) have shown no infections among young inhabitants after the eradication of the vectors. However, live adult worms in nodules and microfilariae in the skin were detected 11 years after the interruption of transmission. Complete elimination of microfilariae from the skin was achieved after 18 years, although onchocerciasis remained the most frequent cause of blindness at this time. This is not surprising because the pathogenesis of the disease indicates that the blinding process must have commenced long after vector eradication when transmission had ceased.

It is clear, however, that transmission has been brought to a halt, although at a price. This price is eternal vigilance and constant surveillance, for reintroduction of *S. neavei* from neighbouring countries and even the establishment of man-biting *S. damnosum* is still a possibility. This is because of the major advances in agriculture and water resources development for irrigation and hydroelectricity, as well as the construction of a network of highways to increase international traffic in that part of Africa.

With all its promise of bringing about control, the use of DDT does have its problems. In the first place it needs to be constantly applied over a considerable expanse of land in order to obviate the dangers of reinvasion. Also, it has to be applied over prolonged periods of 15 years or more, to cover the life span of the adult worms. Furthermore, it has to be applied over numerous and remote rivers and streams, located in difficult and sometimes inaccessible or even dangerous terrains so that the use of aeroplanes and helicopters becomes inevitable. Thus vector control programmes involving DDT require huge investment in money and manpower (trained, experienced, and committed to work under hard conditions), both of which are in short supply in most countries in the endemic areas. Above all, its possible deleterious effect, in the long term, on the environment, and non-target organisms has drawn unfavourable attention to DDT such that, at best, its widescale use over prolonged periods can only be recommended with extreme caution; this is even in spite of the fact that the concentration effective against the larvae is so low as to constitute no danger to other forms of life. Indeed, in the words of Nelson (1978):

> DDT is an undeservedly dirty word and there is no evidence that it is a more serious pollutant of the rivers of Africa than other insecticides. Must we forgo the use of DDT and allow another generation of children to grow up knowing that 20 per cent of them will be blind before the age of 40 years or should we sacrifice a few eagles on the altars of humanity hoping that this will be only a temporary measure whilst new methods of control are developed?

The search for alternative insecticides with greater safety and cost-effectiveness led to the use of Abate (temephos), a phosphorus-based biodegradable insecticide which is effective against *S. damnosum* larvae but is relatively non-toxic to non-target organisms including fish and is safe in drinking water. For these reasons, Abate it the insecticide used in the Volta River onchocerciasis programme.

FILARIASIS IN DOGS

Name of Organism and Disease Caused

Dirofilaria immitis Railliet and Henry, 1911, causes dirofilariasis, filariasis, and heartworm disease.

Hosts and Location

The principle host in the domestic dog; other canids also are good hosts. The domestic cat is seldom infected. There are a number of reports in higher primates, but humans are infrequently infected. Some animals such as the ferret can serve as experimental hosts. The organisms are located as adults in the vena cava, right heart, and pulmonary artery. They are occasionally found in other locations in the dog and especially in foreign hosts.

Distribution and Importance

The organism is cosmopolitan. This worm was originally thought to be found in warm, moist climates such as the southeastern United States, but in the late 1950s data began to accumulate showing that *D. immitis* also occurs in temperate and cold climates. There are enzootic areas near Chicago, Illinois, and Minneapolis, Minnesota; the infection can even become prevalent in semidesert locales such as Grand Junction, Colorado, where irrigation is practiced. It is likely that the organism has extended its range in the United States by increased travel and changing of residence, which Americans seem compelled to do. Surveys in the southeastern United States usually show from 10 to 70 per cent prevalence, whereas that in the northern part of the country is from 1 to 8 per cent. Cats generally have a prevalence of infection less than a tenth that of dogs.

Dirofilariasis is considered to be a major disease of dogs in the southeastern United States and other warm climates.

Morphology

The adult males are from 12 to 200 by 0.8 mm and the females are 250 to 310 by 1 mm, the females are therefore about 1 ft long. Head structures are minimal, but six papillae surround the mouth. The male has 11 pairs of papillae in the anal region and a pair of unequal 300 and 200 μm long, respectively.

Life Cycle

The infective stage, the L_3, develops in mosquitoes, usually in the genus *Aedes,* but other genera are satisfactory vectors in some localities. The L_3s escape from the mouthparts of the mosquito during feeding and enter the wound made by the mosquito. Early in the infection the developing worms occur in subcutaneous, adipose, and muscular tissue. When the worms reach early L_5, at about 90 days postinfection (PI), they begin to locate in the heart. The prepatent period is about 180 days before microfilariae appear in the blood.

Mf are picked up by mosquitoes during blood feeding. They migrate from the gut to the hemocoel and enter cells of the Malpighian tubules, where they become short, sausage-shaped organisms. They then begin to grow and molt twice to reach the L_3 10 to 16 days PI. The L_3s migrate to the head of the mosquito and readily escape when another blood meal is taken.

Diagnosis

The principal means of determining whether an animal is infected is by finding the mf in the circulating blood, but clinical signs are also important in enzootic areas.

Finding the mf is done by drawing a small amount of blood into a weak formalin solution and then looking for them in the sediment after centrifugation. As indicated in the section on *Wuchereria* and *Brugia*, mf can be differentiated from one another, and it is sometimes important to do so. *Dipetalonema reconditum* of the raccoon is most often the filariid that needs to be differentiated from *D. immitis*:

	Length	*Width*	*Tail*
Dirofilaria immitis	286-340 μm	5-7 μm	Straight
Dipetalonema reconditum	269-283 μm	4.3-4.8 μm	Hooked

Other mf landmarks can also be used, but the overall dimensions are most important for these two species.

Signs of infection begin to appear more than 6 months after infection. Dogs lack stamina and may become exhausted or breathe hard after only a short run.

Host-parasite Interactions

Clinical signs appear in dogs as the worms become sexually mature. Give the size, location, and potential number of adult worms, the principal clinical signs involve interference with blood flow. The worms locate in the right heart; when mature they may extend from the vena cava into the right auricle, down into the right ventricle, and out into the pulmonary artery. The impede blood flow and prevent proper closing of the heart valves. There is backpressure from the heart to the liver because the blood does not move through the right heart at a normal rate. The heart works harder in an attempt to empty and also to meet the need to move blood from the heart to all organs of the body.

The results of this study in fluid mechanics are that the liver becomes enlarged as a result of engorgement with blood, the right heart becomes enlarged, and there is insufficient oxygenated blood for the animal to maintain strenuous activity. A frequent pattern is that a dog becomes infected during the summer; it develops a patent (but undetected) infection during the winter at a time when it may be relatively inactive. When spring comes and the exhilaration we all feel wells up, the dog may be turned loose for a romp and suffer from cardiac failure in midgambol. Death or permanent cardiac damage may be the outcome.

There are effects on the host in addition to these mechanical ones. Lesions occur in the blood vessels such as the pulmonary artery, where there may be inflammation are erosions. There mf may also be pathogenic.

There is nocturnal periodicity in the number of mf in the blood of the dog. Peak microfilaremia is from about 8 p.m. to 2 a.m. but periodicity is not as marked as in *Wuchereria bancrofti*.

Treatment

Three aspects to therapy of heartworm infection are as follows:

1. Preventing development of worms in the dog
2. Killing the microfilariae
3. Killing or removing the adult worms

In endemic areas, dogs are given DEC before the start of the mosquito season and for 2 months after it is over This drug prevents the development of the worms that may infect the dog.

If a patent infection develops, the adult worms may be killed with arsenamide or they can be removed surgically in special cases. The drug is administered in divided dosages so that only some of the adults are killed at any one time. If a single large dose of drug were to be given, all of the worms would die at the same time; they would flow downstream and damage the lung so severely that the dog might die.

Surgical removal involves chest surgery in which the worms are extracted by cutting into the major vessels. Special circumstances are required because surgery is both risky and expensive.

The mf may be killed by administering dithiazinine iodine, but care must be taken to ensure that all the mf are not killed at once or a toxic reaction may ensue.

Epidemiology and Control

Although dirofilariasis is more prevalent and more of a problem in warm than in cool climates, it can become a problem in any area where mosquitoes are abundant. The host range is quite broad in the vector, so the probability of its being able to establish in a given area is high. This is confirmed by the cosmopolitan distribution in United States and other countries such as Australia.

In some instances, the incidence of infection has increased dramatically in a locality such a Minneapolis or Cook Country, Illinois, north of Chicago. The increased prevalence of infection in various parts of the United States is probably a result of the high mobility of the U.S. population. People move from one part of the country readily and they travel long distances on vacation. Dogs are taken along and carry their infections back home or to their new homes.

Once established in an area, the infection seems to remain enzootic. The reasons seem to be the long life of the adults, 2 to 3 years, and the high microfilaremia in even moderate infections. Hundreds of mf/mm^3 of blood are common in dirofilariasis. Dogs are seldom protected from mosquitoes and may sleep in unscreened areas near the house. The role of wild canids in perpetuating the infection has not been systematically investigated, but both foxes and coyotes have been found to be infected in enzootic localities.

No public health measures have been taken to reduce the incidence of *D. immitis* infection in dogs, but community mosquito control programs should reduce the probability of infection. The responsibility for control lies with the individual owner. The steps the owner can take are as follows:

1. Protection of the dog from mosquitoes by providing screened quarters or by keeping the dog confined at times of day when mosquitoes most actively seek blood meals, usually at dusk and dawn.
2. Administration of DEC during and after the mosquito season. In areas where mosquitoes are active year round, DEC must be given continuously.
3. Periodic examination of the dog to determine whether it has a patent infection and to provide treatment if necessary.
4. Periodic treatment, usually twice a year, regardless of whether the animal shows signs of infection. Some veterinarians recommend such periodic treatment in areas where the probability of infection is high; the rationale is that the dog probably is infected and the drug will eliminate any infection that might be there. Arsenamide is not without side effects and the risks of treatment should be balanced against its benefits.
5. Clearance of the microfilaremia by dithiazinine. Mf can survive a long time in the blood of an animal after the adult filarioids have been removed. These mf provide a pool of infectious agents to the passed on to another animal, and they may also be somewhat pathogenic to the host. Their removal is therefore a public health measure and gives relief to the individual dog.

The importance of *D. immitis* infections in humans is still poorly known. There have been about 35 case reports of persons infected with what is most likely *D. immitis*. In two there were worms in the heart, but most often they have been found in the lungs and subcutaneous tissues. Dirofilariasis in human is clearly not a major problem, but effects can be moderately severe in infected persons. Prevention of infections in human can be taken care of by control programmes for dogs.

FILARIAL INFECTION OF RUMINANTS AND HORSES

Five genera of filarioid nematodes have been found in domestic herbivores in temperature and warm climates throughout the world. Where systematic studies have been done, the prevalence rate sometimes exceeds 75 per cent. Nearly all species cause at least minor damage to the host and a few cause serious losses in some populations of animals. Significant control progammes do not seem to have been put into practice. Some species that are representative or are of interest because of their impact or populations are the following:

Stephanofilaria stilesi occurs in the skin of the belly of cattle in the United States; it seems to be most common in cattle in the western-United States, where it occurs in up to 90 per cent of animal at slaughter. The skin where the adults and microfilaria are found becomes thickened, inflamed, and leaks blood and serum; such lesions may reach 15 cm (6 in) in diameter. Horn flies, *Heamatobia irritants* (Diptera: Muscoidea), are the intermediate hosts. Losses from this parasite lie principally in trimming of small parts of the hide at slaughter, not a major loss.

A number of species of *Onchocerca* lie ligaments, where they cause inflammation and significant to the host. For example, *Onchocerca cervicalis* occurs in the ligamentum nuchae, the large ligament that inserts at the back of the head and supports it. About 50 per cent of horses in the United States are infected, and upto 80 per cent of horses in both temperate and warm climates in other parts of the world have been found to be infected. The parasite causes inflammation of the ligament and then necrosis of the connective tissue fibers and calcification later in the infection. Vectors are biting midge of the genus *Culicoides* (Diptera, Ceratopogonidae).

The normal hosts of *Elaeophora schneideri* and mule deer (*Odocoileus hemionis*) and white-tailed deer (*Dama virgiana*) in the United States. In the normal host the parasite causes little damage, but there can be considerable damage in domestic sheep and in elk (*Cervus canadensis*). We have discussed some instances in which migratory patterns are altered in a foreign host and a few in which the infection is more severe than in the normal host. *E. schneideri* is a good example of the latter: sheep suffer from a severe allergic dermatitis and elk are often blinded. The infection in sheep was first described by Dr. H.E. Kemper of the USDA in 1938 in sheep in southern Colorado and northern New Mexico that had grazed on pastures above 2000 m (6000 ft) during the summer. It was later found that elk in the Gila National Forest of north central New Mexico were blinded because of infection. An excellent series of studies was done under difficult conditions on the pathogenesis and the life cycle and impact on populations of elk.

LABORATORY MODELS OF FILARIOID INFECTIONS

It is evident that there is a good deal of conjecture and partial knowledge of nearly all aspects of filarioid infections in humans and other animals. Although filarioids may infect many hosts, some have a narrow host range or develop abnormally in other hosts. *W. bancrofti*, for example, develops only in humans; on the other hand, the subperiodic form of *B. malayi* develops fairly readily in primates other than humans and in cats. Thus, we have better descriptions of the course of the disease in *B. malayi* than *W. bancrofti*. An ideal situation is one in which a parasite can be propagated in its normal hosts in the laboratory and studied in a controlled way. *Dirofilaria immitis* has been studied in the dog, but its long prepatent period and the expense of keeping dogs preclude its use as a widespread laboratory model. The best model found to date is *Litosomoides carinii* of the cotton rat; it has been used for a variety of studies on transmission immunity, and other aspects of host-parasite interactions. A great deal has been learned with *L. carinii* but it is of limited usefulness; for example, it can be used as a drug-screening system, but with reservations. Better models would allow knowledge to advance more rapidly.

Chapter—9
Acanthocephala

The Acanthocephala represent a group of worms parasitic in all classes of vertebrates but especially in fish and birds. They show similarities both with the Platyhelminthes and the Nematoda, but no agreement has been reached on their phylogenetic position. They resemble Cestoda in lacking an alimentary canal. The chief diagnostic feature of the group is the presence of an invaginable proboscis armed with hooks form which the common name 'spiny-headed' worm is derived. They range in size from 1.5 mm to over 0.5. The majority of species are small, the largest being *Macracanthorhynchus hirudinaceus* from pigs. The females are nearly always larger than the males. The intermediate hosts, in the few species where these are known, are arthropods, and paratenic hosts frequently occur. In their general anatomy, life history and habits they show a remarkable degree of uniformity. Useful reviews of the biology, ecology, physiology and biochemistry of the group and those of Nicholas (1967, 1973), Rothstein and Nicholas (1969), Bird and Wallace (1969), Pratt (1969) and Crompton, (1970); morphology and taxonomy and dealt with by Hyman (1951) and Yamaguti (1963).

AVAILABLE SPECIES

In Europe, both larval and adult material is fairly readily available. Useful host lists in fish in Great Britain and Ireland have been complied by Chappell and Owen (1969) and Chubb (1970). The following are useful laboratory species:

Table 1. Species of acanthocephala recorded in freshwater fish in Great Britain and Ireland

Species	*Fish host*
Echinorhynchus clavula	eel, perch, roach, grayling, gwyniad, pike, dace, rudd, rudd/bream, trout.
E.salmonis	trout, eel.
E. truttae	trout, grayling, dace, roach, pike, salmon, rainbow, trout, eel.
Neoechinorhynchus rutili	trout, grayling, salmon.
Acanthocephalus anguillae	perch, powan, trout, rudd/bream.
A. lucii	pike, roach, powan, perch, dace, eel, ruffe, trout, bream, rudd, rudd/bream, tench.
Pomphorhynchus laevis	roach, dace, grayling, eel, perch, bream, minnow, pike, rudd, salmon, stone loach, 3– and 10–spined stickleback, tench, trout, rainbow trout, rudd/bream.

Moniliformis dubius

The adult occurs is the laboratory rat, with the cockroach *Periplaneta americana* as the intermediate host; the maturation time in the rat is five to six weeks. Various aspects of the life cycle have been studied by Moore (1946) and Holmes (1961).

Polymorphus minutus

A species common wherever wild fowl or domestic duck, have access to lakes or streams inhabited by the amphipod *Gammarus* spp. One advantage in working with *P. minutus* is that the larva (cystacanths) in gammarids take up a bright orange pigment, with the result that infected specimens may be easily recognized; infections up to 32 per cent have been reported in some areas.

Echinorhynchus truttae

The adults of *E. truttae* occur in brown trout and other fish (with larvae in *Gammarus* spp.) but the larvae are not so obviously pigmented as *P. minutus* (above). Coinvasion of the gammarid by larvae of both *P. minutus* and *E. truttae* is,

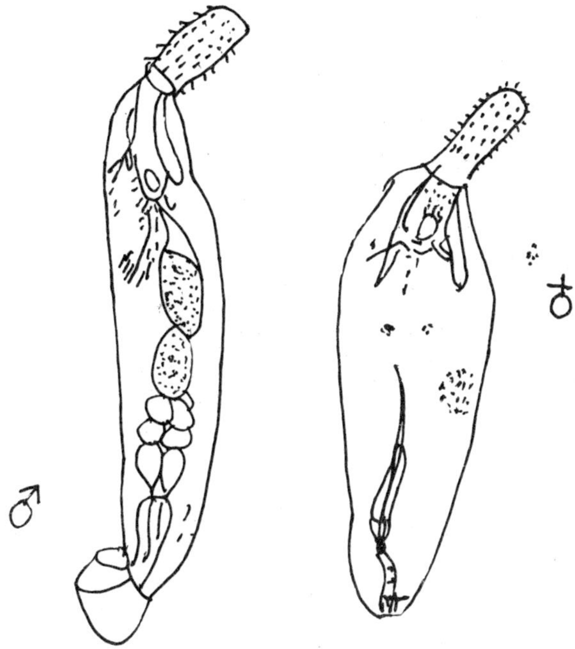

Fig. 9.1 : *Acanthocephalus* sp.—morphology of adult male and female.

in fact, possible. The life cycle of this species has been worked by Awachie (1966).

Acanthocephalus ranae

A relatively common species found in the intestine of the toad *Bufo bufo* its intermediate host is the isopod *Asellus* sp.

Table 2. Maturation of Acanthocephala in the definitive host.

Species	*Final Host*	*Prepatent Period*
Leptorhynchojdes thecatus	rock bass	8 weeks
Echinorhynchus trutae	trout	9–10 weeks
Polymorphus minutus	duck	22–25 days
Moniliformis dubius	rat	5–6 weeks
Macracanthorhynchus hirudinaceus	pig	10–14 weeks

Its mode of attachment to the gut has been studied by Hammond (1967a).

Neoechinorhynchus emydis

A useful species available in the United States in North American freshwater turtles; larvae in freshwater snails. An account of this species has been given by Goodchild (1950) and Hopp (1954).

Macracanthorhynchus hirudinaceus

A large species in wild and domestic pigs with larval stages in scarabaeid beetles. Occasional infections have been reported in man.

Pomphorhynchus laevis

A species in dace in the United Kingdom.

GENERAL MORPHOLOGY

Body

The body consists of an anterior *presoma* with a characteristic spiny proboscis and a neck free from spines, and a posterior *trunk* or *metasoma*. The spines on the proboscis are symmetrically arranged and of diagnostic value. The proboscis (and sometimes the neck) is usually retracted into a proboscis sac by being completely inverted, and the whole presoma is also retractile. Retraction is carried out by means of retractor muscles. Two lateral protrusions from the body wall at the base of the neck are known as the *leminsci*. These are concerned with fluid movement in relation to proboscis activity.

The Body Wall

This consists of a thick tegument which contains a system of *lacunar canals* or *channels* which seem to distribute absorbed food material. It consists essentially of five layers; the outermost or *epicuticle* appears to contain largely mucopolysaccharide, presumably equivalent to the glycocalyx of cestodes. Beneath the epicuticle lies a tough *cuticle* perforated by numerous pores which in turn lead into canals and ducts which comprise the *striped layer*. The latter merges into the fibrous *felt layer* beneath which is a *radial layer* lacking cell walls but containing nuclei, mitochondria, ribosomes, folded plasma membranes, lipid and glycogen. This layer is probably the most metabolically active part of the body wall. Beneath these layers are a thick basement membrane and circular and longitudinal muscles.

The above account is based on light microscope studies supplemented by electron microscopy. A number of species have been examined but (as in cestodes) the body wall shows a remarkable degree of uniformity.

Fig. 9.2 : Fine structure of the body wall of *Polymorphus minutus*.

Representative ultrastructural studies are those on adults: *Polymorphus, minutus, Moniliformis dubius, Acanthocephalus ranae;* acanthor: *M. dubius* egg; *M. dubius*.

Digestive System

This is lacking, as in cestodes. Absorption of materials takes place directly through the body wall.

Excretory System

Lacking except in the order Archiachanthocephala, in which modified protonephridial organs occur. These consist of flame bulbs and ducts which lead into a single duct or bladder.

Nervous System

This is reduced to a ganglion in the wall of the proboscis sac and two nerve trunks which run posteriorly from the brain obliquely to the body wall. In the male, there is a pair of genital ganglia in the peins base.

Nuclear Constancy

The group exhibits the interesting, phenomenon of nuclear constancy or *eutely*. In a given species, the number of nuclei in the organs and tissues attained during larval development remain constant, except in the gonads. Some nuclei in the epidermis of some species fragment into chromatin material.

REPRODUCTIVE SYSTEM

Ligament Sac and Ligament Strand

The ligament sacs are hollow tubes of connective tissue peculiar to the Acanthocephala; these enclose the gonads. In front, they are attached to the proboscis sac or the adjacent body wall, and posteriorly they connect to some part of the reproductive system. The number and arrangement of the ligament sacs varies in the different orders. Thus there are two (a dorsal and a ventral) in the females of the Archi- and Eocanthocephala, but in the males of the same orders only the dorsal sac is present. In the Palaeacanthocephala, there is a single ligament sac in both sexes.

The *ligament strand* is a nucleated strand to which the gonads in both sexes are attached. This strand is believed to represent the endoderm and the body space between this and the body wall may be considered a pseudocoele.

Male System

There are two testes attached to the ligament strand and enclosed in the ligament sac. In the Archiacanthocephala and the Palaeacanthocephala, there occur *cement glands,* six to eight in number, with ducts leading into the common sperm duct. In the Eoacanthocephala, however, the cement glands form a syncytial mass from which a duct enters a *cement reservoir* from which, in turn, ducts enter the sperm duct. The secretion of the cement glands seals the everted male bursa around the female gonopore.

Continuous with the ligament sac is a muscular tube, the genital sheath, which encloses the posterior ends of the sperm ducts, the cement glands and the protonephridial canals when present. There is a short penis which projects into a hemispherical or elongated bursa, a cavity made up of the in-turned body wall with a proximal muscular thickening, the muscular cap. The bursa can be everted to the exterior and is used in holding the female during copulation.

Female System

This shows a number of peculiarities. The ovary or ovaries (two) which are visibly intact only early in life, break up into *ovarian balls,* each consisting of a syncytium from which ovogonia are formed. These soon come to be free; in the Palaeacanthocephala the sac ruptures and they fill the body cavity. The ligament strands lead to a muscular organ termed the *uterine bell.* Posteriorly this opens into a *uterine tube,* bearing two *bell pouches,* which connects to a muscular uterus and a non-muscular vagina opening to the exterior. The vaginal aperture is guarded by a sphincter.

The uterine bell serves as a selective apparatus. Ovic larvae are taken in by the uterine bell and passed through the uterus to the vagina and so to the exterior. Immature eggs are rejected and returned via an aperture in the ventral bell to the dorsal pseudocoele. The selective mechanism in *P. minutus* appears to operate on egg length, most of eggs less than 100 μm being immature. The uterine bell is continuous with the dorsal ligament sac in the Archi- and Eoacanthocephala. In the Palaeacanthocephala, the single dorsal strand connects to the inner side of the uterine bell. In *M. dubius,* an XO- sex determination mechanism appears to operate with 2n=8 (female); 7 (male).

Copulation and Fertilization

The male bursa is used for grasping the female and the penis is inserted into the vagina and sperm discharged. The secretion from the cement glands is then poured over the whole posterior tip of the female and the gonopore securely sealed. The secretion persists for a time as a hard brown 'copulation cap.'

Structure of Eggshell

Four egg membranes have been described in most species, three in others. In *M. dubius* the outermost layers contain keratin, the third is PAS positive and made up of proteinchitin; the innermost layer is proteinaceous. The eggs of some species are extremely resistant, eggs of *P. munutus* retaining their infectivity for 2.5 months at –22°C.

Life Cycle

The life cycles are known in detail for only a few species. The intermediate hosts are insects (larvae, beetles, cockroaches) for those parasitic in land animals, and crustaceans or molluscs for those in aquatic animals. Some developmental stages of *Polymorphus minutus* are shown. The time for hatching of the eggs varies with species, two hours being the earliest reported time for *M. dubius* in the cockroach eggs of *Neoechinorhyncus saginatus* hatch in ostracods within one hour.

Hatching is markedly affected by pH, a pH>7.5 being necessary for optimal result; a molarity > 0.2 is also essential. In the presence of carbon dioxide, however, hatching takes place at a lower pH. A chitinase appears to be involved in the hatching process and bile salts appear to be the chief agent 'activating' larvae in the definitive host.

Classification

Three orders are recognized:

Order 1 *Archiacanthocephala.* Hooks on proboscis either in long rows or in a few circles; spines lacking on trunk; main lacunar vessels dorsal and ventral; both dorsal and ventral ligament sacs present in female; cement glands separate, with eight present in males; in terrestrial hosts, *e.g. Moniliformis dubius.*

Order 2 *Palaeacanthocephala.* Hooks on proboscis usually in long rows; spines on trunk; nuclei in hypodermis usually fragmented; main lacunar vessels lateral; ligament sac in female single; cement glands separate; mostly aquatic hosts, *e.g. Leptorhynchoides thecatus* (in fish; larvae in amphipods).

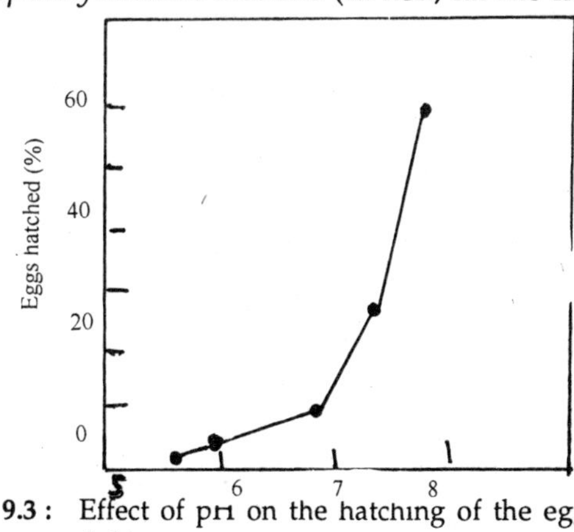

Fig. 9.3 : Effect of pH on the hatching of the eggs of *Moniliformis dubius.*

Order 3 *Eoacanthocephala*. Hooks on proboscis usually in a few small circles; few nuclei (large) in hypodermis; main lacunar vessels dorsal and ventral; dorsal and ventral ligament sacs in female; cement glands syncytial; in aquatic hosts, *e.g.*, *Neoechinorhynchus emydis*.

Several alternative classifications to the above are also in use.

PHYSIOLOGY

General Account

Various aspects of the physiology have been reviewed by Fairbairn (1969), Saz (1969), Rogers (1969), Crompton (1970) and Nicholas (1967). Most physiological work has been carried out on *Moniliformis dubius* and *Polymorphus minutus*.

Since acanthocephalans lack a gut, the uptake of nutrient materials is likely to be similar to that in cestodes. The little work done on this question suggests that amino acids and glucose are absorbed by active transport. The role of the pores in this process is not known but it is likely (but not yet demonstrated) that materials which pass into the pores are more readily absorbed. No unequivocal evidence of pinocytosis has been obtained. Edmonds & Dixon (1966) demonstrated that thorium dioxide and carbon could be taken into the pores, but since the substances were not shown to have passed through the plasma membrane (which lines the pores) absorption could not be said to have been demonstrated.

There is also no evidence available on the relative absorptive capacities of the praesoma and metasoma. The praesoma is in continuous contact with host tissue and it is possible that 'membrane' (=contact) digestion may take place in this region. In this process, the digestive enzymes are believed to be membrane-bound, a spatial arrangement which would permit the products of digestion to be immediately absorbed at the host-parasite interface. In the mesosoma, the same process could (theoretically) occur and in this case, since this region is exposed to intestinal enzymes, absorbed enzymes could be both of intrinsic (*i.e.*, parasite) and extrinsic (*i.e.* host) origins.

Carbohydrate Metabolism

Acanthocephalans contain substantial of glycogen (up to 13 per cent of the dry weight of *Macranthorhynchus hirudinaceus*) and in general have a pronounced carbohydrate metabolism. The glycogen is localized chiefly in the proboscis, the body wall and the muscles. Under host starvation conditions, the glycogen level rapidly falls, in the case of the adult *M. dubius*, to 12 per cent of its initial level within 48 hours. The utilization of glycogen *in vitro* in the larval form is strongly enhanced under anaerobic conditions by carbon dioxide although (rather surprisingly?) the same effects does not occur in the adult.

Few studies on carbohydrate uptake have been made. *Polymorphus* absorbs glucose against a concentration gradient reaching a maximum value once the external level reaches about 2 mg/cm^3. Absorbed glucose would be immediately available for energy purposes and the unused balance for glycogen sysnthesis. *M. dubius* is also able to synthesize glycogen from fructose, mannose and maltose although the uptake mechanisms of these substances have not been investigated. Synthesis of the disaccharide, trehalose by the same species has also been demonstrated.

Table 3. Effects of gas phase on utilization of glycogen in the cystacanths of *Moniliformis dubius*

Gas Phase	*% glycogen utilized*
95% N_2+5% CO_2	58
95% Air+5% CO_2	40
100% N_2	32
Air	34

Like cestodes, Acanthocephalans appear to utilize an Embden-Meyerhof scheme of phosphorylating glycolysis

The chief waste products of carbohydrate metabolism of *M. dubius* are succinic, lactic, formic, and acetic acids together with substantial quantities of ethanol–the latter accounting for some 30 per cent of the original glucose metabolized. It may be noted that ethanol is also produced by cestodes. The pathways to the ethanol or succinate are not known; the formation of succinate may follow that proposed for *Ascaris*.

A significant Krebs cycle has not been demonstrated although it may partly operate in some species. Pyruvate or phosphoenolpyruvate formed from glucose by glycolysis may be carboxylated to form malate which is reduced via fumarate to form succinate. This reduction of fumarate may be an important terminal electron transport system which could result in the re-oxidation of NADH under anaerobic conditions, thus allowing glycolysis and the synthesis of ATP to be maintained. A further similarity with the intermediary metabolism of some cestodes and nematodes is seen in the fixation of carbon dioxide, which in *M. dubius* (incubated in glucose-free media) fixes primarily into malate, succinate and fumarate as well as into aspartic acid, alanine and serine, PEP-carboxykinase appears to be the enzyme chiefly involved in the fixation process. In contrast, Ward and Crompton (1969) found that in a medium containing glucose, CO_2-fixation did not take place! It thus may be that the worms

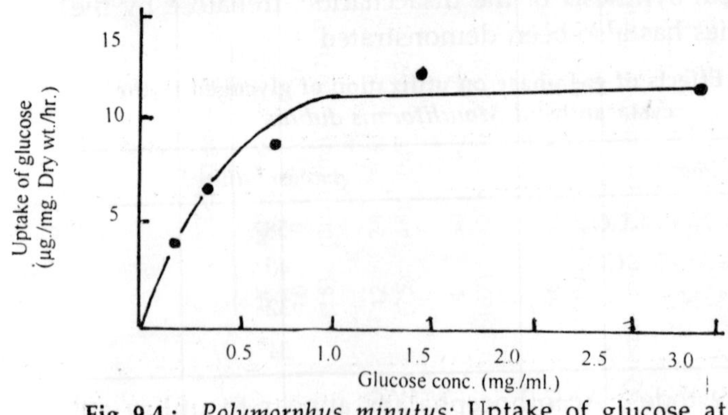

Fig. 9.4 : *Polymorphus minutus:* Uptake of glucose at different concentrations.

only fix carbon dioxide in the absence of glucose. This rather unusual result clearly needs further examination. It would, however, account for the failure of carbon dioxide to stimulate glycogenesis in the presence or absence of glucose as mentioned earlier.

Nitrogen and Lipid Metabolism

Almost nothing to be known of the nitrogen and lipid metabolism of acanthocephalans and this is an areas which calls for further work. It would, for example, be interesting to know whether, like cestodes, acanthocephalans had lost their ability to synthesize lipids and become dependent on the host for their materials.

Knowledge of lipid metabolism has been reviewed by Fairbairn (1969) but only the composition of the lipids appear to be known. The lipid composition of *M. dubius is* σ=4.2 per cent (w.w.), σ=7.2 per cent and that of *M. hirudinaceus* is σ=0.9 per cent and σ=1.7 per cent. It is difficult to account for the fact that the former species contains about four times the lipid in the latter. Fatty acids from C_{10} to C_{20} were identified in both species, but C_{18} acids accounted for more than 70 per cent of the total acids. The phospholipids of both species contained ethanolamine, serine, choline and inositol phospholipids. Of the unsaponifiable lipids, cholesterol accounted for 50-80 per cent.

Chapter—10

Systematic Description of Parasitic Fish Nematodes

Nemathelminths are worms with a cylindrical body, circular in cross-section. Their digestive system consists of three sections: anterior (oesophagus), middle (intestine) and posterior (rectum), ending with the anus.

Instead of containing parenchyma as in plathelminths, the body cavity (pseudocoel) is filled with perivisceral fluid imparting turgor to the body. Parenchyma is reduced to a few mesenchymal cells.

The phylum contains both free-living organisms and parasites of plants and animals. Fish parasites of interest to us are encountered in the class *Nematoda*

Marine fish may be affected by about 300 species of parasitic nematodes. Particularly frequent are larval nematodes of the family *Anisakidae* (subfamily *Anisakinae*).

The anisakids are large nematodes with three lips surrounding the mouth: one dorsal and two ventro-lateral ones, the latter slightly smaller than the former. Some species have lips separated by three smaller interlabia.

The cuticle is clearly striated all over. The straight, cylindrical or slightly widening posteriorly oesophagus consists of two sections: the anterior muscular and posterior glandular part, the so-called ventriculus. The latter connects with the intestine, a straight tube lined with a single layer of columnar epithelium The terminal section of the digestive system is the rectum opening outside via the anus. Three large glandular anal cells are associated with the rectum.

The anisakid nematodes show blunt-ended lobes (caecum and appendix) at the ventriculus-intestine junction. The *Raphidascaris* species have one lobe (appendix) running backwards from the ventriculus; in *Phocanema*, the lobe (caecum) runs forwards from the intestine, while an

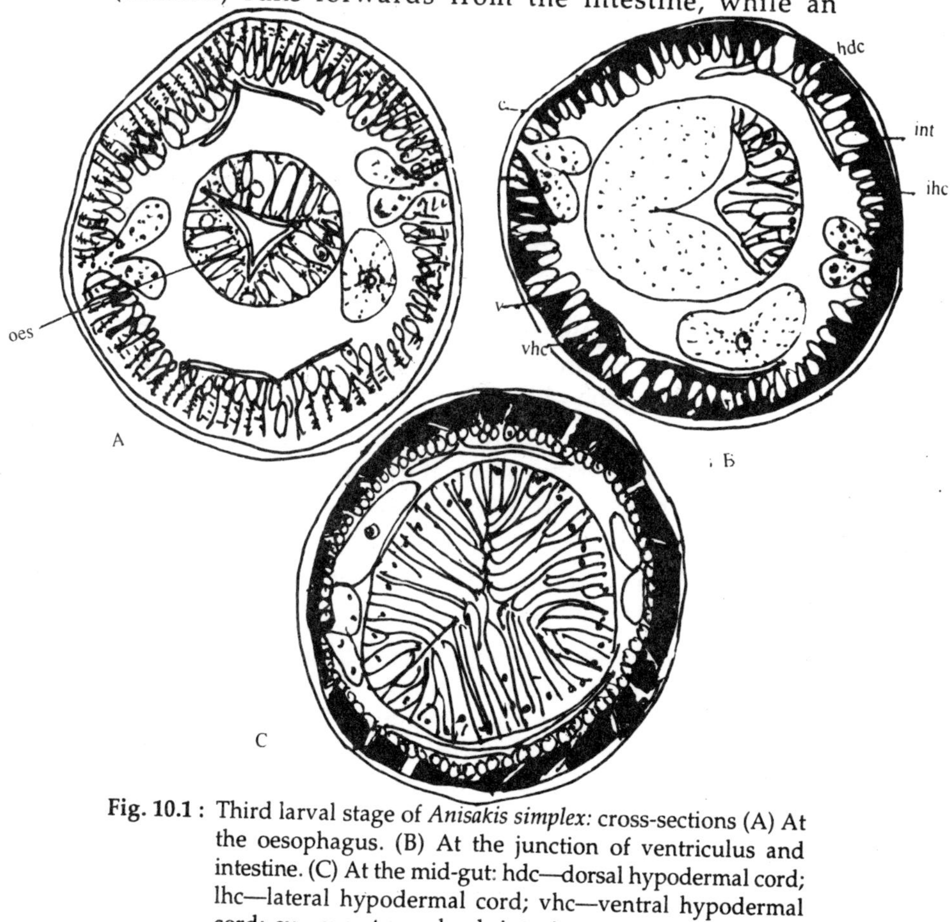

Fig. 10.1 : Third larval stage of *Anisakis simplex:* cross-sections (A) At the oesophagus. (B) At the junction of ventriculus and intestine. (C) At the mid-gut: hdc—dorsal hypodermal cord; lhc—lateral hypodermal cord; vhc—ventral hypodermal cord; ex—excretory gland; int—intestine; c—cuticle m—muscle fibres; oes—oesophagus; v—ventriculus.

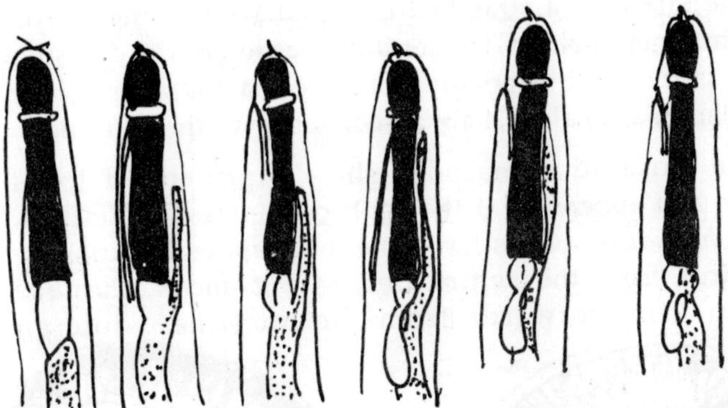

Fig. 10.2 : Diagrammatic presentation of third larval stage of *Anisakinae*. (A) *Anisakis*. (B) *Phocanema*. (C) *Porrocaecum*. (D) *Contracaecum, Phocascaris*, (E) *Thynnascaris*. (F) *Raphidascaris*: stippled part—ventriculus; blackened areas—oesophagus; white areas—intestine.

Contracaecum and *Thynnascaris* one lobe leads from the ventriculus backwards and the other from the intestine forwards. No lobes are found in the genus *Anisakis*, the ventriculus connecting obliquely with the intestine.

The excretory pore is located ventrally: some species have it on the top of the head at the base of the ventro-lateral lips, while in the other species is found near the nervous ring.

The characters mentioned above are of a key importance in genus identification.

The family *Anisakidae* groups species of a major economic importance; some of them are capable of infecting people who happen to consume fish containing live larvae. Those nematodes have very often been described under various names, which might confuse the reader. Therefore in the descriptions the full synonymy of each species is given along with authorities and the recent literature citations.

The genera most important for fisheries are: *Anisakis*, *Contracaecum*, *Thynnascaris*, *Terranova*, *Phocascaris*, and *Raphidascaris*, their larvae being common in marine fishes.

Genus: *Anisakis*

Synonyms: *Stomachus* (Goeze in Zeder, 1800); *Filocapsularia* Deslongchamps, 1824; *Peritrachelius* Deising, 1851; *Conocephalus* Diesing, 1860

The genus contains nematodes whose adults have a mouth surrounded by three lips: one dorsal and two ventro-lateral. Interlabia are absent. Each lip bears a bilobate anterior projection which carries a single dentigerous ridge. The dorsal lip base has two sensory papillae, one papilla each being found on both ventro-lateral lips. The excretory duct opens on top of the head between the ventro-lateral lips. The oesophagus consists of the anterior muscular portion with large glandular cells and the posterior glandular part, the so-called ventriculus, occasionally sigmoid. Ventricular-intestinal blind processes (lobes) absent. The vulva located anteriorly. Male spicules of uneven length; 3 or 4 pairs of postanal caudal papillae present. Adult *Anisakis* nematodes dwell in the stomach and intestine of marine mammals, mostly cetaceans.

The type species is *Anisakis simplex*. In his revision of the genus, Davey (1971) assumes three valid species: *A, simplex, A. typica* (synonym: *A. tursiopis*) occurring in whales of warm and tropical waters between 40°N and 36°S and *A. physeteris* (synonyms: *A. skrjabini; A. oceanica* and *A. brevispicula*). The latter, of a worldwide distribution, occurs in *Physeter catodon* and accidentally in other whales. Davey describes morphological differences between the species of adult forms from their definitive hosts; there are no data on larvae in his work.

Owing to the lack of sufficient information, the following species are treated by Davey as *species inquirendae*: *A. dussumieri, A. insignis, A. schupakovi,* and *A. alexandri*. These species require further studies.

Anisakis Simplex

Synonyms: *A. alata, A. catodontis, A, ivanitzkii; A. kogiae* Johnston and Mawson, 1939; *A kükenthalii; A patagonica, A. pegreffii, A. rosmari; A similis; A. tridentata.*

The species has a relatively long ventriculus, often sigmoid, obliquely connected with the intestine. Spicules uneven; the right spicule shorter than the left one, their length depending on the size of a male. The ratio of right to left spicule length averages 1:1.6, occasionally reaching 1:2. Males and females grow to about 7.2 cm and about 10 cm (sometimes longer), respectively.

A. simplex parasitizes cetaceans and pinnipeds in cold temperate and polar waters. It was recorded from the following marine mammals: cetaceans *Mesoplon bidens, Berardius bairdi, Hyperoodon ampullatus, Physeter catodon, Kogia breviceps, Delphinapterus, leucas, Monodon monoceros, Steno bredanensis, Stenella caeruleoalba, Delphinus delphis, Lagenorphynchus albirostris, L. obscurus., Orcinus orca, Pseudorca crassidens, Globicephala scammoni, Phocaena phocaena, Phocaenoides dalii, Balaenoptera acutorostrata, B. musculus, B. borealis, B. physalus, B. brydei, Megaptera novaeangliae,* pinnipeds

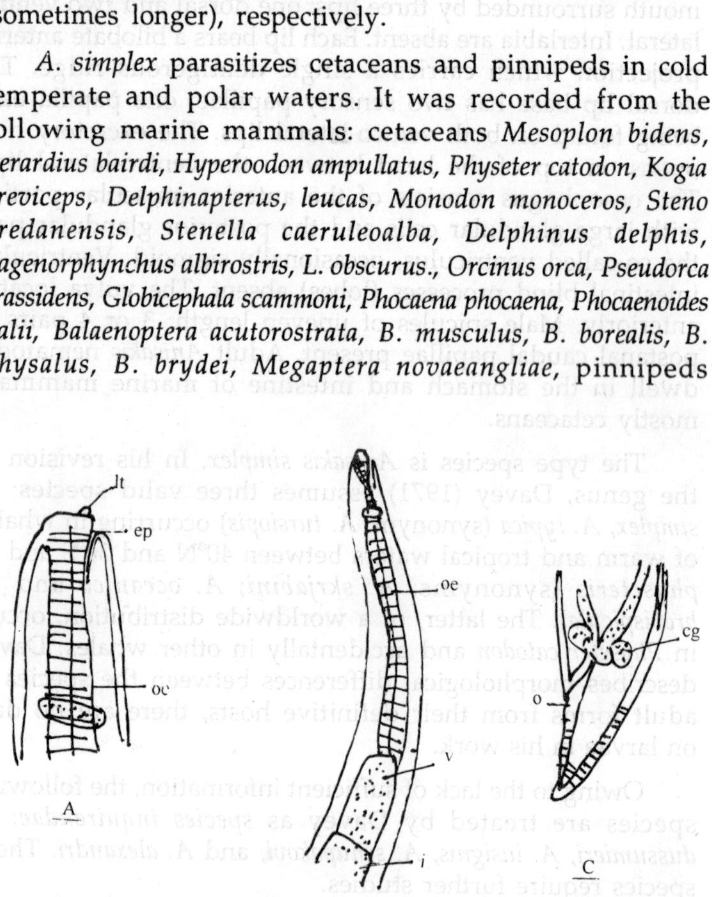

Fig. 10.3 : *Anisakis* sp. type II larva: i—intestine; cg—caudal glands; ep—excretory pore; oe—oesophagus; lt—larval tooth; v—ventriculus; o—excretory pore.

Arctocephalus australis, Zalophus californianus, Eumetopias jubatus, Otaria byroni, Odobenus rosmarus, Pusa hispida, Halichoerus grypus, Hydrurga leptonyx, Monachus monachus, Mirounga leonina, M. angustirostris.

The nematodes settle in the anterior stomach of a cetacean. A heavy invasion (more than 100 nematodes per host) results in numerous crater-like ulcers in the stomach mucosa.

The third larval stage of *A. simplex* occurs coiled as a flat spiral surrounded with a thin delicate cyst in viscera of numerous marine fishes, e.g. herring, mackerel, cod, blue whiting, redfish, Alaska pollock. Moreover, the larvae were recorded in cephalopods (*Todarodes pacificus*). The larvae are worldwide in distribution and are particularly common in the northern hemisphere. According to Japanese literature, the *Anisakis* larvae were recorded off Japan in 132 fish species and in a cephalopod, *Todarodes pacificus*.

Japanese authors distinguish between three types of larvae: *Anisakis* sp. I, *Anisakis* sp. II, and *Anisakis* sp. III. The first corresponds to *Anisakis simplex* from European waters. The second has a much shorter ventriculus connecting perpendicularly with the intestine and conical tail lacking mucron. The larva was originally described by Berland (1961) from *Lampris guttatus* off Norway. Japanese authors found the larva in 32 fish species and two cephalopods (*Todarods pacificus* and *Doryteuthis bleckeri*). Kagei *et al.* (1978) reported an *Anisakis* sp. II invasion in man in Japan.

Anisakis sp. III was recorded only once. This larva has a short ventriculus (like *Anisakis* sp. II) and a tail with a small mucron (like *Anisakis* sp. I).

Anisakis sp. II and III are doubtless larval stages of different, so far, unidentified, species.

The larval *Anisakis simplex,* known commonly as "herring-worm", has been described under various names, most frequent of them being *Anisakis* sp., *Stomachus* sp., and *A. marina.*

They are thin, translucent, slightly pinkish larvae 1.5–2.5 cm long. A larval tooth, slightly eccentrical, is located on top of the head. The end of the body has a contractile process, the so-called mucron with smooth cuticle. Larval *Anisakis* differ from other related nematode genera by the presence of ventriculus between the muscular oesophagus and intestine and by the absence of blind processes, typical of other anisakid nematodes. In the live larva, the ventriculus is clearly visible with the naked eye as a white spot in the cephalic part. The intestine is lined with columnar epithelium. The larva does not feed at this stage; it remains excretory pore is located on top of the head between the ventro-lateral lips.

Spirally coiled larvae usually lie in the peritoneum along the stomach and intestine, under the hepatic membrane, and between the gonads, the latter often showing larval imprints. The larvae are also, although less commonly, found in fish muscles and in the gonad parenchyma. The location of the larvae may depend on the fish species. The most commonly *Anisakis* larvae-affected organ in cod and Alaska pollock in the liver, while in herring the larvae concentrate mainly on

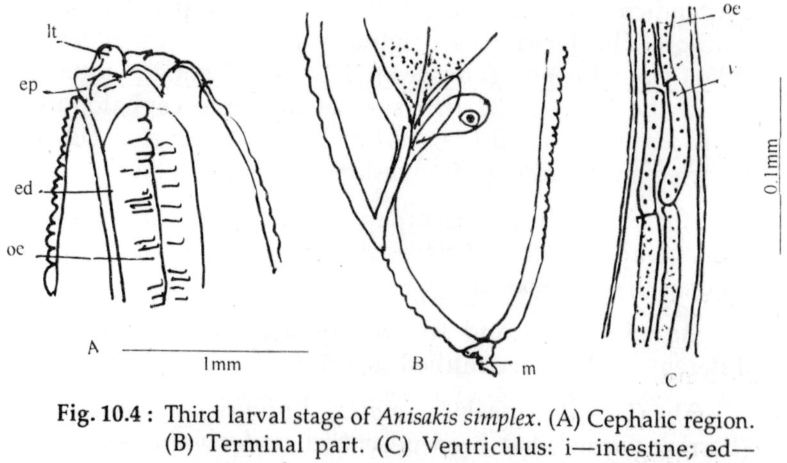

Fig. 10.4 : Third larval stage of *Anisakis simplex*. (A) Cephalic region. (B) Terminal part. (C) Ventriculus: i—intestine; ed—excretory duct; m—mucron; ep—excretory pore; oe—oesophagus; lt—larval tooth; v—ventriculus.

the stomach and along the intestine; muscles are a preferred site in the European blue whiting, whereas in the Baltic garfish, even during a weak invasion with a few larvae per host, they are located immediately underneath the skin and are clearly visible from outside with the naked eye.

The *Anisakis* life cycle proceeds in the following way: the eggs, 41–43×39–42 μm in size, contain 1 or more blastomeres inside a thin translucent shell. The eggs are voided into the water with catacean feces and sink to the bottom where cleavage proceeds and the first larval stage emerges. Some time later the larvae leave the eggs as the second stage, 0.26–0.34 mm long. These larvae are sheltered by a cuticular sheath, a remnant of the first stage ecdysis. The larvae stay in the plankton for some time (under laboratory conditions they remain viable for about 2 weeks) and are eaten by euphausiid crustaceans. In their intestine the larvae lose their sheath and—as typical second stage larvae—move to the crustacean body cavity. These larvae have a distinct larval tooth, a long tapering tail, and a poorly visible intestine. Lips are absent. This stage is infective for fish.

Thus the *Euphausiacea* are the first intermediate hosts for *Anisakis* Oshima (1969) reared the second larval stage *Anisakis* are infected *Euphausia pacifica* and *E. similis*.

Under natural conditions, the *Anisakis* larvae were found in the Barents Sea in *Thysanoessa raschii, Caprella septentrionalis,* and *Hyas araneus*. Smith (1971) found them north of Scotland and off the Faroes in *Th. inermis* and *Th. longicaudata;* he also mentions van Banning's finding of a single larva in *Meganyctiphanes norvegica* in the northern North Sea. Various Japanese authors mention *Th. raschii, Th. longipes,* and *Euphausia pacifica* as the first intermediate hosts in the Pacific. The latter author drew attention to the absence of *Anisakis simplex* in Antarctic *Euphausia superba* and Antarctic marine mammals, which would explain a weak parasitic load in the Antarctic marine mammals, which would explain a weak parasitic load in the antarctic fish.

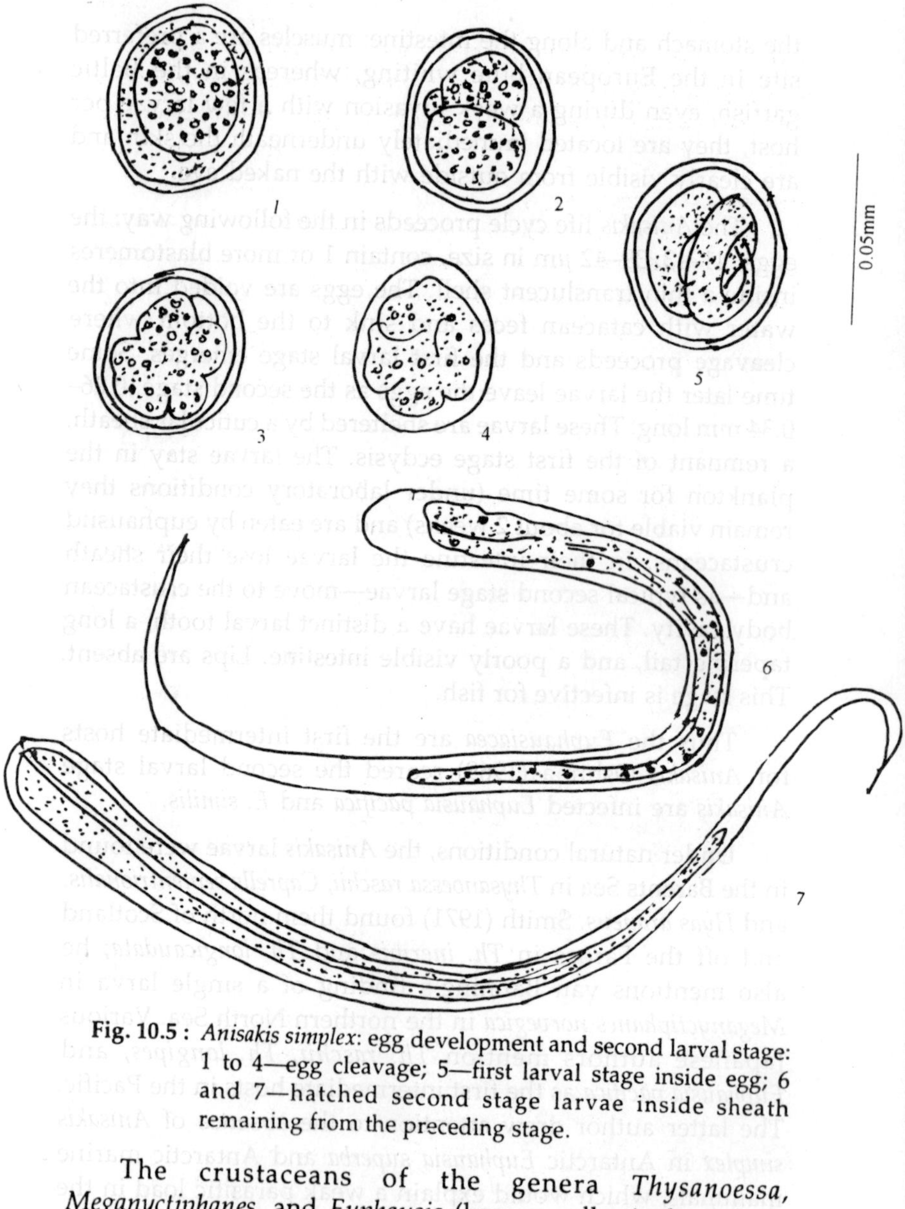

Fig. 10.5 : *Anisakis simplex*: egg development and second larval stage: 1 to 4—egg cleavage; 5—first larval stage inside egg; 6 and 7—hatched second stage larvae inside sheath remaining from the preceding stage.

The crustaceans of the genera *Thysanoessa*, *Meganyctiphanes*, and *Euphausia* (known collectively as krill) are the basic food item of herring, mackerel, and other fish

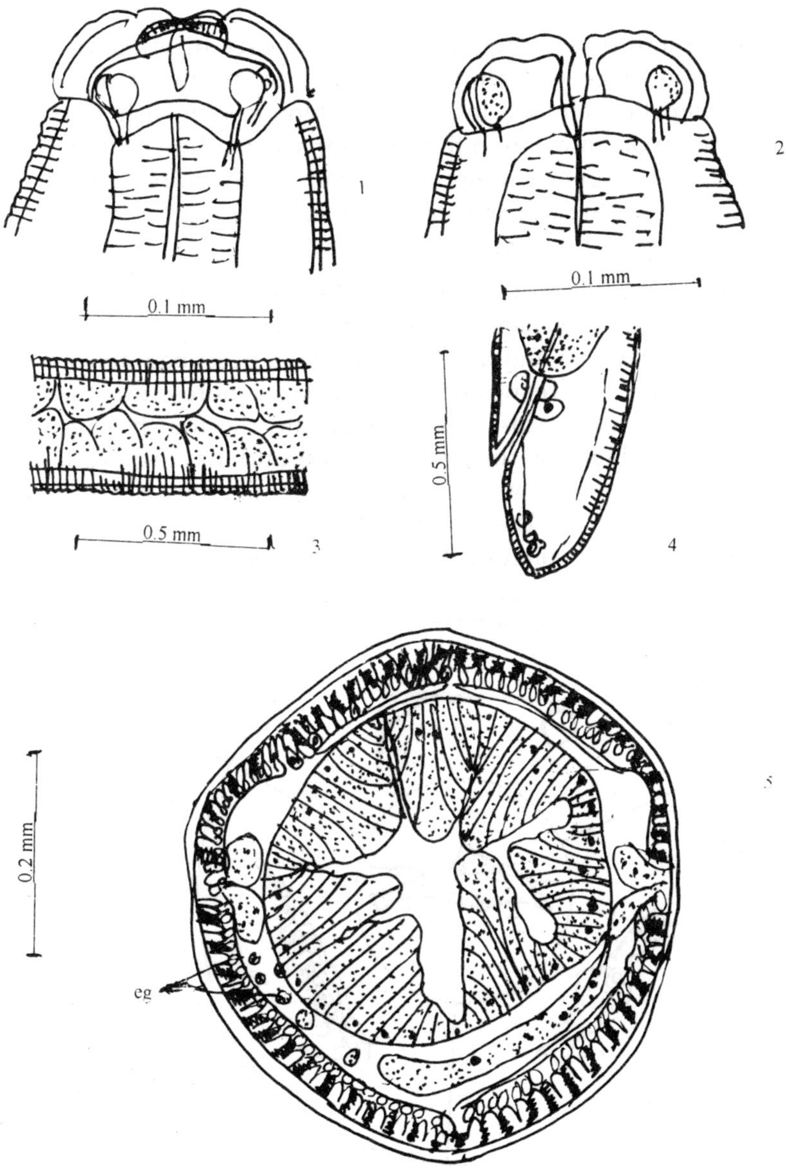

Fig. 10.6 : Fourth larval stage of *Anisakis simplex*: 1—cephalic region, dorsal side; 2—cephaliç region, ventral side; 3—intestine; 4—caudal part; 5—intestine cross-section; eg—emerging gonads.

and cephalopods, the intermediate hosts of *Anisakis*. Moreover, large predatory fish may become infected by eating smaller, already infected plankton-feeding fish. The *Anisakis* larvae move from the predator's stomach to its body cavity and re-encyst there. Thus more and more larvae accumulate in the fish visceral cavity and the fish are at the same time parathenic hosts for *Anisakis*.

The larvae in fish are at their third stage, infective for the definitive hosts, fish-feeding marine mammals: cetaceans (mostly porpoises) and seals. The stage is described above. When in the cetacean stomach, the larva bores into the mucosa and submucosa and moults into the fourth stage (preadult), lacking the larval tooth and having well-developed lips with a characteristics dentigerous ridge at their internal side. The intestine takes up its function. Its triangular cross-section of the third stage becomes star-like;

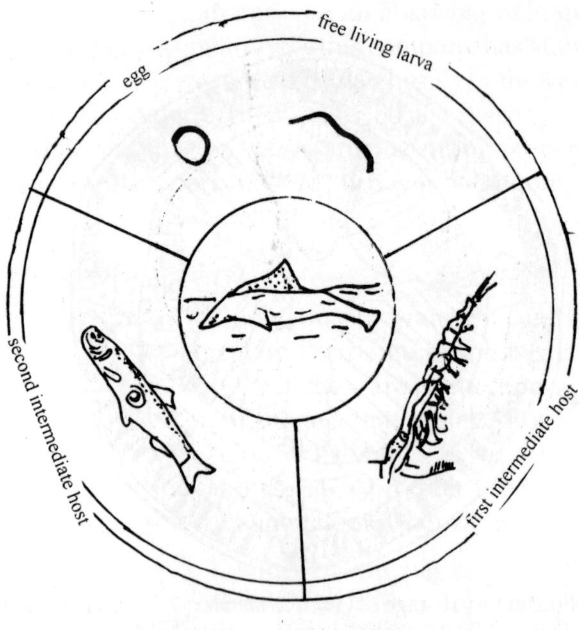

Fig. 10.7 : Life cycle of *Anisakis simplex*.

gonads begin to emerge. The fourth stage larvae leave the cetacean stomach wall and move into the stomach lumen where they moult for the last time into adult males and females (the first stage); these quickly reach sexual maturity and produce numerous eggs.

The development of the larval *Anisakis* from herring was studied, including an *in vitro* culture of the third larval stage, by Banning (1971) and J. Grabda (1976). Fertile males and females were obtained and the larvae could thus be identified to species. They were *Anisakis simplex* both in the North Sea and in the Baltic (J. Grabda, 1976). Other authors attempting to culture the anisakids were able to proceed as for as to the fourth stage, but failed to obtain adults.

As mentioned earlier, the *Anisakis* larvae occur commonly in marine fish worldwide, particularly in the northern hemisphere seas. In the Baltic they were found relatively, recently in herring. First reports were published in 1972. The larvae were also found in the Baltic cod (J. Grabda, 1976) and in garfish (J. Grabda, unpubl.). As shown by J. Grabda (1974), not all the Baltic herring stocks are infected. Larval *Anisakis* occur only in the western herring arriving to the Baltic in autumn from outside to spawn, stay here until spring next year, spawn in March-May, and migrate westwards again. On the other hand, the local Baltic herring stocks such as the spring coastal herring, autumn herring, and spring herring spawning off Sweden are *Anisakis*-free.

The Baltic Sea offers no suitable conditions for the *Anisakis* development. Low salinity precludes the occurrence of euphausiids, the nematode's intermediate host. Porpoises and seals, the definitive hosts, are very scarce so the herring cannot be infected in the Baltic. On the other hand, cod may be infected via herring, its food item. Cod migrations, in the Baltic are very limited in range, the infection taking place mainly in the Pomeranian Bay where most infected western herring concentrate. The *Anisakis* invasion incidence in cod decreases eastwards, accompanied by increasing infestation with *Thynnascaris adunca*.

Most probably, garfish is infected in the North Sea feeding grounds and also via the western herring in the Pomeranian Bay. Information about garfish migrations in the Baltic is scant. Nevertheless, the migrations are presumably very extensive in range, as evidenced by finding the larval *Anisakis* in garfish off SW Finland in the Northern Baltic.

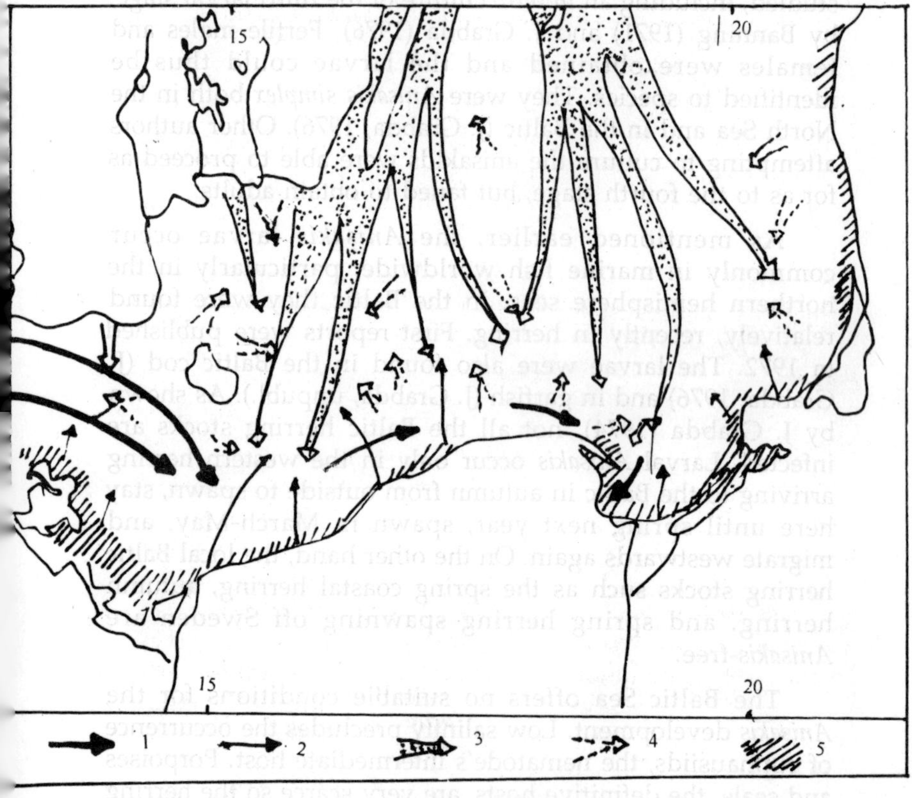

Fig. 10.8 : Herring migrations in the Southern Baltic: 1—spawning migration of infected western herring; 2—feeding migration of Baltic spring herring; 3—feeding migration of Scandinavian spring herring; 4—feeding migration of autumn herring; 5—spring herring spawning grounds in the Southern Baltic.

The *Anisakis* larvae occur not only in marine fish; they were recorded in anadromous species like *Huso dauricus, Oncorhynchus keta* and *O. gorbuscha* during their marine phase. Invasions are known to occur also in freshwater salmonids like brook and rainbow trout, 50 per cent and 26 per cent invasion incidences being recorded in the respective species in Scotland. The infection was presumably spread by marine fish used as fresh food in fish farms. A particularly simple invasion pathway in trout maricultures should be taken into account; for this reason, only frozen food should be used.

Similarly to *Thynnascaris adunca,* the larval *Anisakis* seriously damage fish liver. Most larvae settle beneath the connective tissue of the hepatic membrane; some of them, however migrate deeper into the parenchyma, thus destroying the tissue around them. A liver, heavily affected by *Anisakis,* may be smaller than a healthy one, as observed in cod and redfish.

A strong invasion of adults are fourth larval stage produces stomach ulceration, sometimes very extensive, in cetaceans.

The third larval stage of *Anisakis* is invasive for man.

Genus: *Thynnascaris*

Synonym: *Contracaecum* Railliet and Henry, 1912

The genus consists of nematodes with three lips and three interlabia. The ventriculus has a posteriorly directed lobe, while the intestinal lobe is stretched anteriorly. The lips lack teeth. The excretory pore is located at the level of the nervous ring. Both the third and fourth, as well as adults dwell in fish.

Thynnascaris adunca

Synonym: *Contracaecum aduncum*

The adult form inhabits intestines of cod, eelpout, and numerous other marine predatory fish. The nematode is very common in the northern hemisphere oceans and seas (Atlantic and Pacific as well as the North, White, Baltic, Bering and other seas). It is particularly common in the Baltic eelpout (*Zoarces viviparus*). *Thynnascaris* may attack other animals besides fish: high numbers of these nematodes were encountered in prawns *Pandalus borealis*.

Adult males and females of *Th. adunca* reach about 20 mm and about 36 mm in length, respectively. Apart from generic characters, the species shows fine spines on the tail.

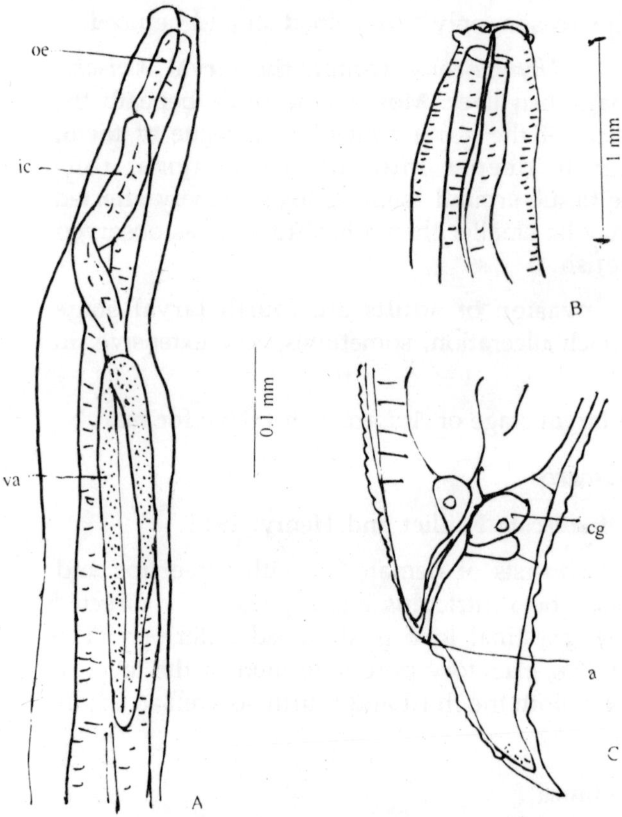

Fig. 10.9 : *Thynnascaris adunca:* third larval stage from Baltic cod. (A) and (B) Anterior section of the body. (C) Caudal region: cg—caudal glands; i—intestine; oe—oesophagus; a—anus; ic—intestinal caecum; va—ventricular appendix.

The spines appear in the fourth larval stage and are sometimes visible through the cuticle in the third stage larvae.

The life cycle of *Th. adunca* requires two intermediate hosts. The first are copepods, as demonstrated by Markowski (1937) who experimentally infected *Acartia bifilosa* and *Eurytemora affinis*. Copepods are infected after consuming eggs, released to the water by the affected fishes with their feces. The eggs hatch into larvae; these moult and pierce through the copepod intestinal wall to get to the body cavity. Copepods in turn infect planktivorous fish (clupeids, flatfish, and others) in which the development proceeds further. Once in fish, the larvae pierce the intestine and get into the body cavity where they moult again and, as the third stage, encyst on the intestinal walls, between the pyloric caeca, and/or in the liver. As opposed to *Anisakis*, the *Yh. adunca* larva do not coil; they are also smaller than the *Anisakis simplex* larvae and reach 6–16 mm, seldom more.

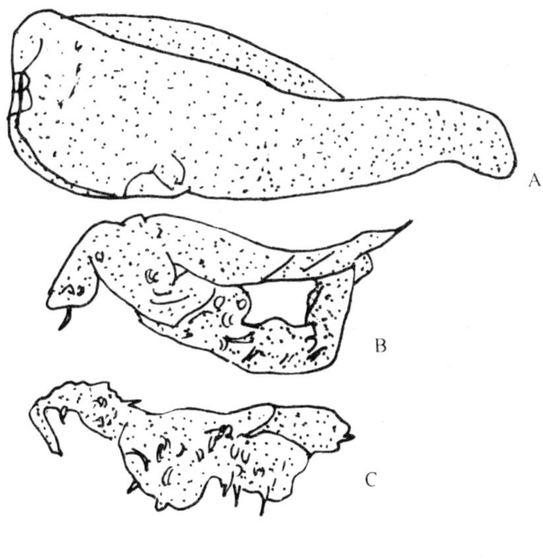

Fig. 10.10 : (A) Healthy cod liver. (B) and (C) Pathologic changes in cod liver affected by the presence of *Thynnascaris adunca* larvae. The livers were obtained from cod individuals of equal size.

The larvae remain encysted for about a year and perish unless consumed by the definitive host: cod, eelpout, or other predatory fish.

A particularly strong *Th. adunca* invasion was recorded in the Baltic cod. Individual livers were often observed to contain more than 100 larvae. The infection of cod in the Southern Baltic affected 55 per cent of the population as observed in 1974–1975, the invasion extent increasing eastward.

When badly affected, the fish lose weight and their condition gets worse. An affected liver is much smaller than a healthy one; in extreme cases it may be completely destroyed. Condition coefficient of the cod containing 200–300 larvae is 0,629 as opposed to 0.941 of a healthy individual. The liver fat content of a strongly infected fish can drop from 57 per cent to 14.5 per cent.

An invasion of *Th. adunca* larva can be fatal for newly hatched fish. As shown experimentally by Rosenthal (in Sindermann, 1970), high mortality rate was typical as early as in the first month of culture in the larval nematode-infected juvenile herring. The infection was observed to affect even the 9–13 mm long herring larvae. The swiftly moving nematodes were visible in the larval herring abdomens; during the first 6–8 days after infection the nematodes grew to equal the larval fish intestine length. By coiling around the intestine, the larval *Th. adunca* stop the peristaltic movements, as a result of which the fish larva affected cannot feed and must die. Under natural conditions, a considerable proportion of fish fry doubtless dies out as a result of the invasion. The strength of the invasion increases with fish age; a 55% invasion incidence was recorded in the 30–40 cm long fish, the incidence increasing to 97% in the 80–100 cm long ones.

According to Soviet authors, the vitamin A content in the larval *Thynnascaris*-infected cod does not change, so the liver can still be used for cod liver oil production.

Th. adunca is relatively harmless for man, as all developmental stage occur in invertebrates and fish, i.e. poikilotherms, there are opinions that the nematode may affect people by producing in condition known as eosinophilous granuloma.

Intestines of marine fish may be inhibited by other *Thynnascaris* species, the identification of which requires elaborate studies, rendered more difficult by the fact that the species are described by various authors under different names.

Genus: *Phocascaris*

The genus includes large nematodes with three lips. Depending on the species, interlabia are present as such, may be reduced to small papillae, or absent altogether. A row of fine spines may occur on the inside of lips. The excretory pore is located on the top of the head, at the base of the ventro-lateral lips. The ventriculus in short, with a backward-directed lobe; the intestinal lobe is directed forward.

Berland (1963) assigns 6 species to the genus. All of them, both the adults and the fourth stage larvae (preadults) parasitize pinnipeds. The third larval stage lives in marine fish.

Phocascaris cytophorae Berland, 1963

Synonym: *Contracaecum* sp. Berland, 1961

This large, brown-coloured nematode lives in the stomach and doudenum of the seal *Cystophora cristata*. Males and females grow to 100 mm and 135 mm respectively. Adults have three lips; interlabia absent, small papillae occurring between the lips (the papillae are absent in the third larval stage). Lips are lacking teeth. The excretory pore is located on top of the head, below the ventro-lateral lips. The intestinal lobe originating at the junction of the ventriculus and the intestine runs backwards.

The third stage larvae are mostly 12–13 mm (occasionally 35 mm) long; they occur in marine fishes. Berland (1961) found them located on the liver, between pyloric caeca, and visceral membrane of cod (*Gadus morhua*) and lump-sucker (*Cyclopterus lumpus*) off Norway.

The life cycle is incompletely known. The second stage larvae are presumed to live in marine invertebrates.

Phocascaris osculata (Rudolphi, 1802) Berland, 1963

Synonym: *Contracaecum osculatum* (Rudolphi, 1802) Baylis, 1920

Adults live in the stomach and intestine of *Phoca hispida, Ph. hispida, Ph. groenlandica, Erignathus barbatus,* and other seals of the northern hemisphere. The nematodes are particularly common off the east coast of Greenland and around the British Isles. The parasites have small ventriculus with a somewhat clavate lobe directed posteriorly; the intestinal caecum is much longer and anteriorly directed. The mouth is surrounded by three lips and three interlabia; the excretory pore located between the ventro-lateral lips.

Adult females are up to 52.0 mm long and 1.6–2.1 mm thick; males grow to 40–53 mm in length and are 0.9–1.5 mm thick. According to Berland, females reach 90 and males 70 mm in length. They usually occur in swarms; the anterior portion of the body is embedded in creater-like depressions in the intestinal mucosa of seals.

The development of *Ph. osculata* is incomplatetely known. Small marine crustaceans are probably the first intermediate hosts. Herpacticoid copepods (*Idya furcata* and *Amphiascus similis*) were experimentally infected.

Eggs, kept *in vitro,* hatch into the second larval stage 400–600 μm in length, surrounded by a cuticular sheath, a moult vestige of the first stage developed in the egg. The sheath is lost inside copepods. The second intermediate hosts are marine fishes in which the third stage larvae develop.

Ph. osculata larva are very similar to those of *Ph. cystophorae* and may even be misidentified. Therefore, to invoke the opinion of Berland (1963), only the generic name (*Phocascaris* sp.) should be applied to these larvae.

The third stage larvae of *Phocascaris* were also recorded in the Baltic herring and other fish species caught off Finland.

Of other *Phocascaris* species, adult forms of the following are known: *Ph. phocae* parasitizing *Phoca groenlandica, Ph. netsiki* from *Phoca hispida, Ph. radiata* recorded in three species of Antarctic seals, and *Ph. turgida* recorded only in the Hawaiian seal *Monachus schauinslandi.*

Genus: *Contracaecum*

Lips of these nematodes dentigerous ridges. Interlabia well-developed; the excretory pore located at the base of the ventro–lateral lips. The ventriculus small, reduced, its lobe directed backward. The intestine provided with a lobe stretching anteriorly. The larvae resemble these of *Thynnascaris*. Adults live in piscivorous birds. The genus numbers about 30 species. The type of species is *Contracaecum spiculigerum* Railliet and Henry, 1912 of a world-wide distribution; it occurs in the intestine of many birds, e.g. cormorants, auk, heron, seagulls, pelican, gannet, and others. The third stage larvae were found in fish.

The life cycle of this and other species are poorly known. According to Soviet authors, larval forms of *C. spiculigerum* are 15–24 mm long. *C. siluriglandis* larvae occurring in the body cavity and muscles of catfish, bream, ide, and other fishes of the Aral and Caspian Seas and the central Asian lakes. Dubinin infected pelicans and cormorants with these larvae and obtained sexually mature *C. spiculigerum*. Adult nematodes grow to 40 mm (males) and 85 mm (females).

Genus: *Phocanema*[1]

These nematodes have three large muscular bilobate lips, each with a dentigerous ridge inside. No interlabia. The excretory pore located at the base of ventral lip; the ventriculus lacks appendix. A single large, backward-directed process is observed at the ventriculus-intestine junction. The nematodes parasitize marine pinnipeds.

The types species is *Ph. decipiens.*

Phocanema decipiens (Krabbe, 1878) Myers, 1959

Synonyms: *Terranova decipiens* (Krabbe, 1878), Baylis, 1916;

Porrocaecum decipiens (Krabbe, 1878) Baylis, 1929

Sexually mature individuals of the species dwell in the intestine of marine mammals: *Arctocephalus gazella, Balaenoptera acutorostrata, Callorhinus ursinus, Enhydra lutris, Eumetopias grypus,* and other living in the North Sea, North Atlantic, North Pacific and their adjacent seas as well as in Arctic and Antarctic waters. The nematodes are large: males reach 52 mm and females 64 mm in length.

The third stage larvae occur in muscles and liver of numerous marine fishes such as cod, Alaska pollock, eelpout,

---

1. The generic name *Phocanema* was introduced by Myers (1959). The type species is *Ph. decipiens,* its larvae—known as "codworms"—being recently observed to cause anisakiasis in people. The genetic names *Terranova* and *Porrocaecum* have been frequently used in the literature to denote the latter species, which Myers qualifies as erroneous. In her opinion, the genera *Terranova* and *Porrocaecum* group nematodes parasitizing sharks and piscivorous birds, respectively, while the *Phocanema* nematodes occur in the digestive system of marine mammals. The third stage larvae are very similar in all three genera; they are virtually identical in *Phocanema* and *Terranova*. On the other hand, the *Porrocaecum* larvae have the intestinal caecum directed forward, as in *Phocanema* and *Terrocaecum* but they are additionally equipped with interlabia, the excretory pore being located at the level of the nervous ring.

flatfish. When eaten together with their host by a larger fish the larvae retain viability, migrate to the new host's muscles and remain invasive (similarly to *Anisakis*). Thus the fish are their parathenic hosts; the larval invasion increases with fish growth and age.

Ph. decipiens larvae are large (up to 58 mm length), reddish to dark brown in colour. They lie irregularly coiled in fish muscles, the muscles becoming hyperemic around the parasites. The larvae are particularly frequent in cod (hence the vernacular "codworm"), the muscles of a badly affected fish becoming loose and spongy.

Demersal fish on the shelf are most vulnerable, the pelagic and open sea fishes being much less affected. The fish infection with the large *Phocanema* larva is a much more serious problem for the fish market than any other larval anisakid invasion. The other anisakid larvae are much smaller, whitish, and therefore less conspicuous.

The first intermediate hosts for *Ph. decipiens* are marine invertebrates. Myers (1960) found the larvae to dwell in annelids (*Nereis*), amphipods, isopods, mysids, and other crustaceans as well as in molluscs are experimentally infected echinoderms. Particularly frequent are the larvae in the fish in the vicinity of seal colonies. *Phocanema* larva made up 75% of all the nematodes found in fishes of the New found land fishing grounds.

Recently, a number of reports on human infections with *Phocanema decipiens* larvae have appeared.

Genus: *Raphidascaris*

Synonyms: *Hysterothylacium* Ward and Magath, 1916; *Ichthyascaris* Wu, 1949.

Nematodes of this genus have a small ventriculus with a lobe directed backward. The intestine lacks lobes. The smooth is surrounded by three lips without dentigerous ridges. No interlabia. The excretory pore is located at the level of the nervous ring.

The genus groups 12 species described so far from freshwater, marine, and migratory fish. The best known species is freshwater *R. acus*, typical of the genus and described below as an example.

Raphidascaris acus (Bloch, 1779) Railliet and Henry, 1915

Synonyms: *Ascaris acus* Bloch, 1779; *Ascaris piscicola* Linstow, 1878

The parasite occurs in inland water of Europe, Asia, and North America; it is very common in Poland, particularly in the northern part of the country. Adults inhibit the intestine of pike (principal host) and other predatory fishes like pike-perch, perch, eel, catfish, stickleback, burbot, and others. The species was also recorded from *Clupea alosa* and *Belone belone*. Females grow to 45 mm; males are slightly smaller, up to 33 mm in length.

The larvae occur in numerous fish species, mostly cyprinids. The life cycle involves the first intermediate host, a chironomid larva (mainly *Chironomus chironomus*) (Engashev, 1964) or—more seldom—an oligochaete (*Tubificidae* and *Naididae*).

The *R. acus* larvae leave eggs in the water; when swallowed by a chironomid larva or an oligochaete they migrate to the host's body cavity and moult into the second larval stage which, after some time (14 days in Engashev's experiment), becomes infective for fish. Cyprinids, when feeding on benthic invertebrates including the parasite's intermediate hosts, become infected in the process. When in fish, the larvae make their way to the body cavity, moult again, the encyst on the intestine, mesentery, liver, or gonads. Particularly strong invasions were observed in perch. The larvae are very small (up to 3 mm in length).

The definitive hosts are predatory fish. The development may presumably proceed without involving a cyprinid because not only the adult nematodes, but also larvae, encysted in the abdominal cavity, were found in predatory fishes.

A strong invasion of adults causes inflammation of the intestinal mucosa. Kozicka (1953) observed pathological changes in the intestinal mucosa of a pike containing 93 nematodes.

R. acus larvae produce many more pathological effects. A strong *R. acus* invasion in bream was described by Osmanov (1953) in an Amu-Daria delta lake; 100% of the fish population were infected with an intensity of up to 1035 larvae per host. The fishes were badly emaciated. Deep lesions were seen in the liver and intestine, and gonads were atrophied. The changes observed in late October led to the death of the total bream population in winter.

A similar invasion was observed by J. Grabda (1968) in the Wegielsztyn Lake near Wegorzewo. The bream were badly emaciated and had sharp dorsal ridges and concave sides, "board-like" as described by local fisherman. The fish viscera were "seeded" with *R. acus* larval cysts. The cysts were also located in the liver which additionally contained numerous unencysted larvae, and in the gonads. The abdominal cavity contained a bloody effusion; the visceral and lateral peritoneal membranes were spongy, pink-red in colour, and strongly infiltrated. The liver was rusty, pink or yellowish with a dark-green hue, congested in place. The intestine was empty, its mucosa very loose and covered with large amounts of dense mucus.

The epizootic affected large bream, 33.0/39.5 to 42.0/ 62.5 cm (body length/total length) in size. The affected fish weight ranged within 575–1140 g, while the healthy bream of about 60 cm in length may weight 5–6 kg. Bream juveniles were only slightly affected. The infestation doubtless commences when bream switches to benthic feeding, which usually happens in the third year of life. In the course of fish growth the larvae accumulate in the body, which generally results in a pathologic condition.

In the sea, the genus *Raphidascaris* is represented by *R. anchoviellae*, occurring in the intestine of *Anchoviella epsera* and *Menidia menidia* off North America, and *R. chirocentri* found in *Chirocentrus dorab, Zeus faber,* and *Chaunax fimbriatus* from the Sea of Japan and East China Sea.

The *Raphidascaris* nematodes are harmless for man.

The subfamily *Anisakinae* contains 24 genera. The adults occur in the intestine of fishes, reptiles, amphibians, birds, and mammals. The developmental cycle is complex are usually involves an intermediate host and parathenic ones, sometimes a whole series of them. Some species, e.g. *Anisakis simplex, Phocanema decipiens,* and *Phocascaris adunca* cause fish-borne anisakiasis in man, the third larval stage being infective. For this reason, a correct identification of the larvae is very important.

It is very difficult, and virtually impossible at times, to identify the larvae to species. The identification is corroborated or refuted when adult forms are cultured *in vitro* from larvae, the adult characters, forming a basis of a correct species diagnosis. Usually the generic name derived from certain genus-specific characters is used to denote a larva. The basic genetic criterion is the structure of digestive system, namely the presence of ventriculus, the presence or otherwise of lobes at the ventriculus-intestine junction, and the excretory pore location. Sometimes the latter is difficult to see, in which case the larva is classified *sensu lato* as a type, e.g. *Phocanema*-type larva.

The *Anisakinae* diagnostic characters are presented diagrammatically.

Family: *Philometridae*

The family contains filamentous, large, viviparous, nematodes with the amphidelic reproductive system. Adult females have no vulva, so the larvae liberate themselves by breaking through the uterus and body wall. As opposed to the *Anisakidae,* instead of lips they have 6–8 papillae arranged in two rings around the mouth. The intestine is a straight dark-brown tube. Females lack and anus, the intestine terminating blindly, its end attached to the body wall.

The philometrids live in the fish body cavity, between fin rays, in the subcutaneous tissue, and in gonads. The parasites are of a considerable economic importance. A mass occurrence of the gonad-dwelling species may cause a partial or total castration of the host, thus reducing the affected fish population. Larvae migrating within the fish body are particularly harmful for juvenile fish. Those species settling in skin cause ulcers, the affected individuals being rejected from commercial or game catches.

The list of genera includes *Philometra, Philometroides, Philonema, Thwaitia* and others. They affect freshwater and marine fishes.

The philometrid nematodes are still relatively poorly known. Males are unknown in many cases, the species descriptions being based on females only. Similarly, life cycles of numerous species are unknown. They have been followed in certain freshwater species only, e.g. *Philometra ovata* described by Molnar (1966) and *Philometrodes sanguinea* described by Wierzbicki (1960). The cycles are presumably similar in other species of the family, therefore the cycles of the two species mentioned above are described, as examples, below.

Philometra ovata (Zeder, 1803)

The females are large (up to 12 cm long and 0.8–1.2 mm in diameter), while males are very small (1.7–2.3 mm long and 0.045–0.053 mm thick). Mature females, red in colour, dwell in the body cavity of European cyprinids, mainly bream and roach. They are frequent in the *Ligula*-infected fish.

According to Molnar (1966) females filled with enormous amounts of larvae leave the fish and get to the water where the difference in the osmotic pressure causes them to break down and release the first larval stage. 0.350–0.510 mm long and 0.012–0.022 mm in diameter. The larvae are eaten by copepods (*Cyclops strenuus, Macrocyclops, albidus, Acanthocyclops viridis*). Once is a copepod body, the larvae make their way to its body cavity and stay there for 15 days, after which time the copepod dies. The fishes become infected by feeding on the infected copepods. The larvae push through the fish intestine and settle at first in the swim bladder serous, membrane where they develop further. After the copulation, likely to occur there, females move to the

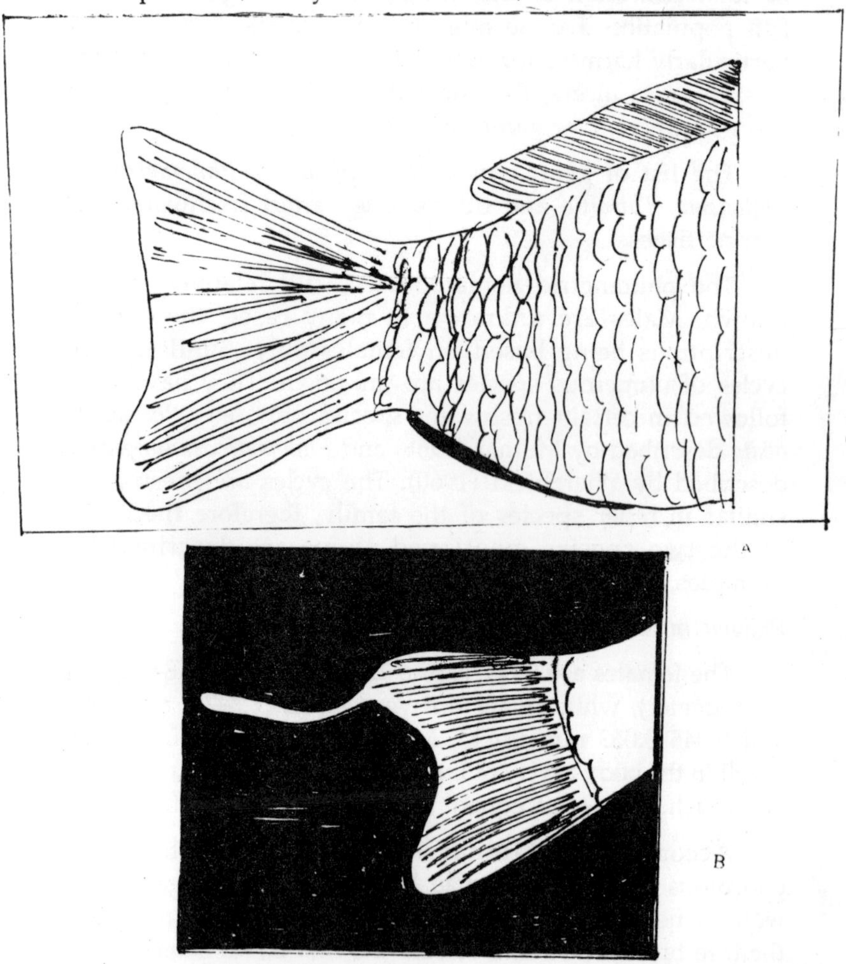

Fig. 10.11 : *Philometroides sanguinea.* (A) Four females in caudal fin of crucian carp. (B) Female moving out of the caudal fin of crucian carp, before larvae are released.

body cavity and mature, while males as a rule die in the swim bladder.

Similar is the life cycle of *Philometroides sanguinea* (Rudolphi, 1819) (synonym: *Philometra sanguinea*) described by Wiezbicki (1960). The cycle spans one year. Adult, 42 cm long females (1.0 mm diameter), blood-red in colour, occur in the interradial tissue of the crucial carp. They appear in the fish in September and stay throughout the winter and spring till the end of May or early June at which time, filled with larvae, they get out from the fins. Their body walls break and release 0.37–0.40 mm long larvae to the water. The larvae live freely for up to 9 days, their infective ability being retained for up to 5 days.

Further development proceeds in the body of large predatory copepods: *Cyclops, Mesocyclops,* and *Macrocyclops*. The larvae migrate from the copepod intestine to the body cavity and reach a stage infective for fish. Having been consumed together with their copepod hosts by a fish, the larvae move to the neighbourhood of the swim bladder to grow, differentiate into males and females, and copulate. Subsequently, the males (2.35–3.30 mm long) die, while females migrate to the fish fins.

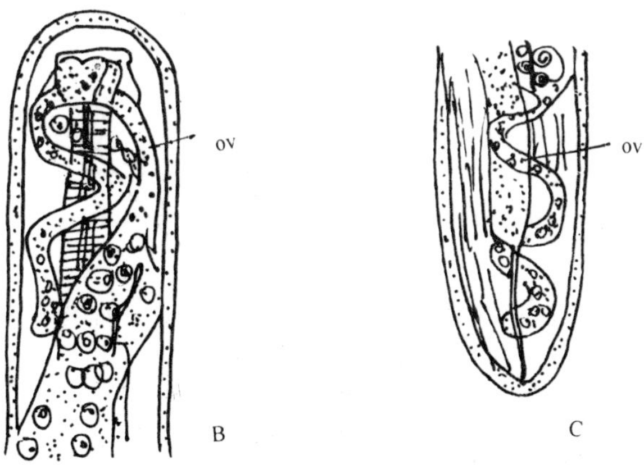

Fig. 10.12 : *Philometra globiceps*. (A) Ovary of nematode-infected blue fish (B) Female anterior region. (C) Female posterior end: ov—ovary.

Philometra americana Kuitunen-Ekbaum, 1933

The nematode parasitizes in fins and subcutaneous tissue of marine fishes. Kuitunen-Ekbaum (1933) found it in *Platichthys stellatus, Lepidopsetta bilineata,* and many other Pacific fish off the Washington State and British Columbia coasts. It is common parasite, frequent in game and commercial catches. Large, red-coloured females form repulsive-looking nodes on the fish body surface, the affected fish being lost for market and processing.

The life cycle is not known, but it is supposed to proceed in a way similar to that of other *Philometra* species. The fish is infected by copepods, the first intermediate hosts.

Philometra globiceps (Rudolphi, 1819)

Large filamentous red-coloured nematodes with dark-brown intestine. Females reach 60 mm in length and may grow even to 200 mm, according to some authors, and attain about 1 mm in thickness. Males are much smaller (up to 6 mm long and 0.08 mm thick.) The nematodes live in the fish body cavity and gonads. They were found in *Blennius phycis, Uranoscopus scaber. Labotes surinamensis, Pomatomus saltatrix* and other fishes of the Adriatic. Mediterranean, and Black Seas as well as in the Atlantic (off Bermuda). The *Uranoscopus scaber* production in the Black Sea was found to be infected in 35.1–86.5 per cent, the parasite burden amounting to 2–9 nematodes per host. Invasion intensity may occasionally be very high, resulting in a total or partial destruction of the gonads.

The life cycle of the species is known in fragments only. Presumably, during fish spawning, the nematode body walls break up and the larvae get to the water with the fish sexual products. The larvae are subsequently eaten by copepods which in turn infect the fish. At 22–24°C, the larvae of *Philometra* survive unencysted for 8–9 days. The life cycle is presumed to span a year.

Philometra lateolabracis (Yamaguti, 1935)

The nematode dwells in ovaries and oviducts of various marine fish species such as *Lateolabrax japonicus, Scomberomorus chinensis, Scianena schlegeli, Otholitus ruber* and others off Japan, Pakistan and India. Dark-red filamentous females, up to 10 cm long and 0.53 mm thick, and filled with larvae, 0.47 mm in length and a maximum of 0.019 mm in width. Males are unknown.

Philometra pellucida (Jägerskiöld, 1893)

The nematode lives in testes of *Tetrodon stellatus, T. hispidum, Dactylopagrus macropterus, Belone liura,* and *Sebasticus marmoratus* off Australia and Japan. Dark-red coloured females reach 45–85 mm in length and are 1 mm thick. Males are unknown.